THE M. & E. HANDBOOK SERIES

PROPERTIES OF MATERIALS

C. V. Y. CHONG
B.Sc.(Lond.), Ph.D.(Lond.), C.Chem., F.R.I.C.

Principal Lecturer in Materials Science and Head of Division of Construction Science and Materials, Polytechnic of the South Bank SW 8.

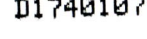

D1740107

MACDONALD AND EVANS

MACDONALD & EVANS LTD.
Estover, Plymouth PL6 7PZ

First published 1977

©

MACDONALD AND EVANS LIMITED
1977

ISBN: 0 7121 1659 1

HANDBOOK *Conditions of Sale*

Printed in Great Britain by Richard Clay (The Chaucer Press), Ltd., Bungay, Suffolk

PREFACE

It has long been realised that there is a need for a book on Properties of Materials to be used by students of various disciplines and at various levels of work, from H.N.C. to first-year degree. In an attempt to achieve this objective, this volume has been prepared to provide an introduction to the properties and uses of the more common materials used in the construction industry. It is intended for use by students preparing for examinations in Properties of Materials or Materials Science in degree, professional, H.N.D. and H.N.C. courses in Architecture, Building, Engineering and Surveying, and others concerned with the construction industry. Students preparing for O.N.C. and City and Guilds Examinations may find some chapters (*e.g.* on cement and concrete) suitable for their needs. The references at the end of each chapter should provide the student with a source of further information on any desired topic. The progress tests have been designed to test and monitor the student's knowledge of the appropriate subject matter.

The author wishes to acknowledge his thanks to the following examination boards for their kind permission to reproduce examination papers:

The Society of Engineers (S.O.E.)
The Council of Engineering Institutions (C.E.I.)
The Institute of Building (I.O.B.)
The Polytechnic of the South Bank (P.S.B.)
Brixton School of Building (now Vauxhall College of Building and Further Education) (B.S.B.)

The author also wishes to thank his colleagues, particularly Mr Arnold Cleveland B.Sc., Hon. F.I.C.W., who has patiently read through the manuscript and provided useful suggestions and advice for improvement; the publishers, Macdonald & Evans, for their help and encouragement; the various publishers, authors and organisations for the information quoted throughout the book; and others who may have been inadvertently omitted.

ACKNOWLEDGMENTS

1. *American Society for Testing and Materials.*
   Data from ASTM C 91–71
2. *British Standards Institution.*
   Information, data and extracts from various British Standards Specifications
3. *Medical and Technical Publishing Co. Ltd.*
   Building Limes
   Gypsum plasters
   Building Stones
   Major applications of plastics
   (*The Construction Industry Handbook*, 1973 edition)
4. *Director, BRE, DOE.*
   Typical composition of Portland and other cements
   (J. W. Figg and S. R. Bowden, *Analysis of concrete*, 1971)
5. *Sir Isaac Pitman and Sons Ltd.*
   Typical compound composition of White Portland cement
   (A. M. Neville, *Properties of concrete*, 1973 edition)
6. *Applied Science Publishers Ltd.*
   Total Heats of hydration of cements
   (D. F. Orchard, Concrete Technology, Vol. 1, 1973 edition)
7. *Institute of Structural Engineers*
   and
   Cement and Concrete Association.
   Workability for different purposes
   (*Report on concrete practice*, 1963 edition, now replaced by publication No 48.037 *Concrete practice*)
8. *Society of Engineers Inc.*
   Summary of some admixtures
   (C. V. Y. Chong, *Concrete Admixtures*, J. Soc. of Eng., Vol. LXIII No. 4: 1972)

9. *University of Nottingham*
   Department of Civil Engineering
   and
   The Refined Bitumen Association.
      Practical bitumen—aggregate mixes
      Effect of composition on mix proportions
      (Bituminous Materials for Flexible Pavements, Residential course at the University of Nottingham, April 1970)
10. *The Institute of Mechanical Engineers.*
      Comparative properties of rubbers
      (Handbook, *Engineering Materials and Methods*)
11. *Directors of ICI, Paints Division.*
      Formulations for: Low build "Alloprene" / inert plasticiser paint (ICI Mond 11D 182/1)
         Inert high build paints for brush application (ICI Mond 11G 216/1)
         Zinc dust brush primer (ICI Mond 11A1)
12. *Chapman and Hall Ltd.*
      Formulations for:
         Wood primer
         Alkyd gloss finish
         Cheap oleoresinous gloss finish
         Emulsion paint based on PVAc
      (G. P. A. Turner, *Introduction to Paint Chemistry*, 1967 edition)
13. *J. H. de Bussy.*
      Formulations for:
         Air-drying Building Undercoat, white
         (Cray Valley Products Ltd. formulation 1339A)
         Thick white brush coating
         (ICI Mond 11G28)
         White fungicide gloss paint
         Zinc dust brush primer (ICI Mond 11A1)
         Antifouling composition (Insoluble matrix type)
         Brilliant white jelly gloss paint
         (Cray Valley Products Ltd. formulation)

      (Materials and Technology, Vol. V., *Natural organic products*, 1972 edition)

14. *Addison-Wesley Publishing Company Inc.*
   Summary of galvanic cells
   (L. H. Van Vlack, *Materials Science for Engineers,*
   1970 edition)
15. *Edward Arnold (Publishers) Ltd.*
   Comparison of hardness values
   Comparison and properties of irons
   (E. C. Rollason, *Metallurgy for Engineers,* 1973 edition)
   Total heats of hydration of cements
   (F. M. Lea, *The Chemistry of Cement and Concrete,* 1970
   edition)

NOTICE TO LECTURERS

Many lecturers are now using HANDBOOKS as working texts
to save time otherwise wasted by students in protracted note-
taking. The purpose of the series is to meet practical teaching
requirements as far as possible, and lecturers are cordially invited
to forward comments or criticisms to the Publishers for consideration.

P. W. D. REDMOND
General Editor

CONTENTS

APPENDIXES

LIST OF ILLUSTRATIONS

LIST OF TABLES

CHAPTER I

BUILDING LIMES

INTRODUCTION

Lime is the product obtained when limestone is strongly heated or calcined:

$$CaCO_3 \xrightarrow[\text{(\sim900° C)}]{\text{heat}} CaO + CO_2 \quad . \quad . \quad . \quad (1)$$

Limestone   Lime   carbon
(or marble   (quick-   dioxide
or chalk)   lime)

Lime, in this context, means quicklime (calcium oxide, CaO). However, the word "lime" is often loosely used also to mean slaked lime (calcium hydroxide, $Ca(OH)_2$), or even limestone (calcium carbonate, $CaCO_3$).

The process of "burning" limestone was known to the ancient Greeks and Romans who made mortar by slaking the lime and mixing it with sand.

Nowadays, the decomposition (equation (1)) by heat is carried out in shaft and rotary kilns (similar to those used in the manufacture of cement—*see* Chapter IV) to produce limes whose composition and properties depend on the composition of the raw material (*e.g.* limestone, etc.) and on the efficiency of the "burning" process.

CLASSIFICATION

Limes can be conveniently classified as:

(*a*) Non-hydraulic.

       (*i*) High calcium (or fat) limes.
       (*ii*) Magnesian limes.

(*b*) Hydraulic.

1. Non-hydraulic limes. These are so called because they will not set or harden under water.

(*a*) High calcium (or fat) limes. These are the products of calcination of fairly pure limestones containing 95–97 per cent calcium oxide. They react vigorously with water ("lime-slaking" process) to form calcium hydroxide, resulting in about 20 per cent expansion:

$$CaO + H_2O \rightarrow Ca(OH)_2 + \Delta H \text{ (1165 kJ/kg)} \quad (2)$$

Quick-   Water   Calcium   Heat evolved
lime   hydroxide or
                    slaked lime

Under carefully controlled conditions, the quicklime can be hydrated with a stoichiometric amount of water (*i.e.* just sufficient water as calculated from equation (2)) to form a dry powder known as *dry hydrate*. When an excess of water is used in the slaking process, a suspension of calcium hydroxide in water known as *milk of lime* is obtained.

High calcium limes are used mainly in mortars, renderings and plasters. They gain in strength very slowly and very little, due to evaporation of absorbed water and to carbonation from the atmospheric carbon dioxide:

$$Ca(OH)_2 + CO_2 \xrightarrow{\text{slow}} CaCO_3 + H_2O \quad (2(a))$$

Slaked lime   Calcium
(or calcium   carbonate
hydroxide)

(*b*) Magnesian limes. These are the calcination products of dolomitic limestones which contain a magnesia (MgO) content greater than 5 per cent (in some cases, even over 40 per cent):

$$\begin{array}{c}
\left.\begin{array}{c} CaCO_3 \\ MgCO_3 \end{array}\right\} \xrightarrow{\text{heat}} \begin{array}{c} CaO \\ MgO \end{array} + \begin{array}{c} CO_2 \\ CO_2 \end{array}
\end{array}$$

atmospheric CO_2 (carbonation) H_2O(slaking) $\quad (2(b))$

$$\begin{array}{c} Ca(OH)_2 \\ Mg(OH)_2 \end{array}$$

2. Hydraulic limes. These are so called because they can set or harden under water. The hydraulicity is due mainly to the presence of dicalcium silicate ($2\ CaO.SiO_2$) and partly to the presence of aluminates. Hydraulic limes are therefore products of calcination of limestones containing certain amounts of clayey matter (which contains silica (SiO_2) and alumina (Al_2O_3)). Depending on the composition of the raw material (limestone) and on the temperature of decomposition (above the range 900°–1200°C), a series of limes can be obtained, varying from fat limes (containing less than 2 per cent of alumina and silica), lean or semi-hydraulic limes (containing more than 5 per cent of alumina and silica) to eminently hydraulic limes (containing up to 50 per cent of alumina and silica). The rate of hydration (slaking) and heat evolution decreases with the increase of the alumina and silica content. With the finely-ground eminently hydraulic limes the reaction with water is slow but it finally sets into a hard mass (similar to a set cement).

PROPERTIES

BS 890: 1972 (Specifications for Building limes) covers the following types (excluding eminently hydraulic limes):

3. Hydrated lime (powder): (a) High-calcium lime (white lime); (b) High-calcium by-product lime; (c) Semi-hydraulic lime (grey lime); (d) Magnesian lime.

4. Quick lime: (a) High-calcium lime (white lime); (b) Semi-hydraulic lime (grey lime); (c) Magnesian lime.

5. Lime putty: (a) High-calcium lime (white lime) putty; (b) High-calcium by-product lime putty; (c) Semi-hydraulic lime (grey lime) putty; (d) Magnesian lime putty.

Properties and specifications are summarised in Table I.

USES

Building limes are used as:

6. Mortars for brickwork, blockwork and masonry, and for plastering and rendering (see Table II).

Adding lime to a cement/sand mortar improves workability and plasticity, cohesiveness and bonding characteristics. These, in turn, enhance such properties as durability and resistance to rain penetration.

TABLE I: PROPERTIES AND SPECIFICATIONS OF BUILDING LIMES

Type	Hydrated lime	Quicklime	Lime putty
Composition	Mainly $Ca(OH)_2$	Mainly CaO ($+$ some Ca Silicates and Ca Aluminates in Semi-hydraulic limes)	Mainly $Ca(OH)_2$ (+ some Ca Aluminates in Semi-hydraulic limes)
Manufacture	1. By hydration of quicklime with minimal water to produce a *dry powder* 2. By treatment of Calcium Carbide (CaC_2) with water, as a by-product in the manufacture of acetylene, followed by drying	By burning limestone or chalk at $1000°$–1100 °C to produce *lumps*	1. By treatment of hydrated lime or quicklime with excess water to produce a *putty* 2. By treatment of CaC_2 with excess water, as a by-product in the manufacture of acetylene, followed by other processes to produce a putty
Fineness	Total residue on 90 μm sieve $<6\%$		Total residue on 90 μm sieve $<6\%$
Colour	White to light buff	White to light buff	White to light buff
Bulk density	550–600 kg/m³	800–1000 kg/m³	1500 kg/m³
Chemical Analysis: $CaO + MgO$	$>65\%$ ($>60\%$ for semi-hydraulic lime)	$>85\%$ ($>70\%$ for semi-hydraulic lime)	$>65\%$ ($>60\%$ for semi-hydraulic lime)
MgO (except for Magnesian limes)	$<4\%$ (Magnesian lime $>4\%$)	$<5\%$ (Magnesian lime $>5\%$)	$<4\%$ (Magnesian lime $>4\%$)
CO_2	$<6\%$	$<6\%$	$<6\%$
Insoluble material	$<1\%$	$<3\%$	$<1\%$
Soluble Silica (SiO_2)	$>5\%$ (semi-hydraulic lime)	$>6\%$ (semi-hydraulic lime)	$>5\%$ (semi-hydraulic lime)

Compressive strength	1:3 lime:sand can attain a compressive strength of up to 7 N/mm² in six months (Compare: Portland Cement 28 N/mm²)		
Transverse strength	Modulus of rupture: For semi-hydraulic lime only: 0·07–0·21 N/mm² at 28 days		
Rheological properties	>12 bumps on flow table	>14	>14
BS test for workability	To reach a spread of 190 mm		
Soundness (Lechatelier test)	<10 mm (except High-Calcium by-product lime)	—	<10 mm (except High-Calcium by-product lime)
Effect of chemicals	Semi-hydraulic limes are attacked by Sulphates		
Reaction with water	No heat evolved	Considerable heat evolved (heat of hydration 1165 kJ/kg)	No heat evolved
Durability	Lime/sand mortars are not very resistant to frost when first used, since rate of hardening is slow. Subsequent behaviour is very good except where mortar remains very wet. Plastering mixes made of lime usually require addition of Portland Cement or gypsum plaster to provide adequate early strength.		

TABLE II: USES OF BUILDING LIMES

Uses	Main desirable properties	Typical mixes
Mortars (for brickwork, block-work and masonry)	Good workability and plasticity Good bonding characteristics and adequate strength Good durability and resistance to frost Pleasing appearance	Nominal mixes by volume—1:2:9 and 1:1:6 (cement[1]: lime: sand) Pigments may be added
Internal Plasters	Good workability, strength and durability Good bonding and cohesive properties Pleasing appearance	Depending on background material and required finish Nominal mixes by volume— *Undercoats* 1:2:9 and 1:1:6 (cement[1]: lime:sand) 1:3:9 (gypsum plaster: lime:sand) *Finishes* 0–1:1:0–1 (gypsum plaster: lime:sand)
External Rendering	Pleasing appearance Good workability and plasticity Good bonding properties, strength and durability	As for bricklaying mortars Pigments optional
Soil stabilisation	Improved strength and load-bearing capacities	Lime added: 2–6% by weight of soil

[1] Cement used: Ordinary Portland, Portland Blastfurnace or Sulphate-resisting Portland. *Never* use High-alumina cement.

7. Stabilisers for soil and other materials (*see* Table II).
Building limes, especially in the hydrated form, are added to a
wet clayey soil in amounts varying from 2 to 6 per cent by
weight of soil in order to improve strength and bearing
capacities of soil. This arises from a chemical reaction between
lime and clay minerals in soil resulting in an immediate
reduction in plasticity of the clayey soil. Also used in lime-
established pfa (pulverised fuel ash) and pfa–lime–soil mixtures.

Relevant BS specifications and codes of practice:

BS 4721: Ready-mixed lime:sand for mortar.
CP 111: Structural recommendations for load-bearing
walls.
CP 211: Internal plastering.
CP 221: External rendered finishes.

PROGRESS TEST 1

1. How is lime obtained? (p. 1)
2. How may limes be classified? (1–2)
3. With the aid of chemical equations describe the setting and
hardening process of a building lime. (1–2)
4. What is meant by the term "hydraulicity"? (2)
5. What are the main causes of hydraulicity in building
limes? (2)
6. Summarise the preparation and properties of building limes.
(Table I)

EXAMINATION QUESTION

1. Discuss the uses of lime in building.

FURTHER READING

Lime in Building, The British Quarrying and Slag Federation,
February 1974.

CHAPTER II

GYPSUM PLASTERS

INTRODUCTION

Gypsum plasters are basically calcium sulphate derived from partial or complete dehydration of gypsum ($CaSO_4.2H_2O$).

Probably they were one of the earliest cementing materials known, being used in Europe in prehistoric times and by the ancient Egyptians for building purposes, for plastic arts and for the conservation of the dead.

Gypsum occurs naturally in different forms (Gypsum rock, Alabaster, Selenite, Anhydrite, Gypsite, etc.) as solid deposits in various parts of the world, the main deposits in the U.K. being in Nottinghamshire, Staffordshire and Derbyshire.

TYPES OF GYPSUM PLASTER

Gypsum is chemically calcium sulphate with two molecules of water of crystallisation and is represented by the chemical formula $CaSO_4.2H_2O$. Depending on the conditions of heating, the following changes occur:

	SG	Crystal structure
$CaSO_4.2H_2O$ (Dihydrate)	2·32	monoclinic

↓ about 150 °C

$CaSO_4.\frac{1}{2}H_2O$ (Hemihydrate or Plaster of Paris)	2·75	monoclinic

↓ 190–200 °C

$\gamma CaSO_4$ (Soluble anhydrite)	2·61	hexagonal

↓ about 600 °C

$\beta CaSO_4$ (Insoluble anhydrite or dead-burnt gypsum)	2·96	rhombic

↓ 1100–1200 °C

$CaO + SO_3$ (decomposition products: calcium oxide and sulphur trioxide)

According to BS 1191 gypsum building plasters are grouped in four classes:

class *A*—hemihydrate gypsum plaster (Plaster of Paris).
class *B*—retarded hemihydrate gypsum plaster.
class *C*—anhydrous gypsum plaster.
class *D*—Keene's plaster.

1. Plaster of Paris. This is a form of hemihydrate $(CaSO_4 . \frac{1}{2}H_2O)$ without any retarder of set. It is manufactured in several grades for various industrial uses. Finely-ground gypsum is usually heated to about 150–160°C in shallow open iron vessels, when 75 per cent of its water of crystallisation is driven off. The product is a hemihydrate commercially known as Plaster of Paris. This, on adding water, sets very rapidly— setting time varying from 5 to 20 min.—accompanied by the evolution of heat and expansion of material:

$$CaSO_4 . \tfrac{1}{2}H_2O + \text{water} \xrightarrow{\text{very rapid set}} CaSO_4 . 2H_2O$$
$$\text{Hemihydrate} \hspace{5cm} \text{Dihydrate}$$

Plaster of Paris, due to its rapid set, is not normally used as a plastering material, but may be used for patching and other purposes where a rapid set is required.

2. Retarded hemihydrate gypsum plaster. This contains an added retarder of set, such as 0·1 per cent keratin. This enables the setting time to be extended by an hour or more. It may be used as a finishing coat for plasterboards or as an undercoat on various surfaces, *e.g.* concrete, bricks, breeze slabs, etc.

BS 1191 subdivides class *B* into 2 types:

(*a*) *Undercoat plaster* (plaster for use with sand).

1. Browning plaster.
2. Metal lathing plaster.

(*b*) *Final coat plaster.*

1. Finish plaster.
2. Board finish plaster.

On adding water, setting is at first delayed due to the protein retarder added, but then becomes rapid, being converted to the dihydrate.

Some typical trade names of class *B* plasters include Adamant, Belpite, Battleaxe, Cretestone, Gothite, Gypstone, Murite, Paristone, Pytho, Sirifix, Thistle.

3. Anhydrous gypsum plaster. This is known as the moderately-burnt anhydrous gypsum plaster which is obtained by heating gypsum at 190–220°C to drive off all the water of crystallisation. This anhydrous form (soluble anhydrite) is very hygroscopic, readily absorbing water vapour to form the hemihydrate. An accelerator is normally added to speed up the set. On adding water, setting is slow but continuous. It is therefore suitable for finishing only.

Typical trade names of class *C* plasters include Cilastone, Sirapite, Statite, Victorite, Xelite.

4. Keene's plaster. This is the hard-burnt or dead-burnt anhydrous gypsum plaster obtained by heating gypsum to a temperature of about 600°C to give a relatively inert product (insoluble anhydrite). An addition of 0·5–1·0 per cent potash alum or potassium sulphate accelerator is added to accelerate the rate of crystallisation to produce setting when water is added. A slow but continuous set occurs, which makes it suitable for finishing only. Both classes *C* and *D* plasters hydrate or set into the dihydrate form.

Typical trade names of the class include Keene's cement, Parian cement, Superite.

However, if calcination of gypsum is carried out at a much higher temperature (*e.g.* 1100–1200°C), some decomposition into calcium oxide and sulphur trioxide occurs and a solid solution of CaO in $CaSO_4$ is formed. This is the basis of the Estrich Gips, the German flooring plaster, which has a very slow set.

PROPERTIES

Gypsum—calcium sulphate dihydrate—forms colourless crystals of specific gravity 2·3 and a hardness of 1·5–2 on the Moh's hardness scale (*see* Appendix II). Its solubility in water is about 2 g $CaSO_4$/litre at 20 °C and that of the calcined gypsum is 8 g $CaSO_4$/litre (soluble anhydrite) and 2 g $CaSO_4$/litre (insoluble anhydrite) at 20 °C.

It has a relatively good resistance against chemicals such as acids, alkalis and oxidising agents, although in contact with water it softens and consequently loses strength.

The setting and hardening of calcined gypsum is a two-stage process: the hydration and dissolution of calcined gypsum in

TABLE III: PROPERTIES OF GYPSUM BUILDING PLASTERS

Bulk density	A and B: 700–950 kg/m³
	C and D: 800–900 kg/m³
	Browning: 500–700 kg/m³
	Metal lathing and bonding ⎫ 500–600 kg/m³
	Finish ⎭
Porosity	According to water/plaster ratio.
	Normally high for A, fairly high for B
Water-absorption	High (especially A), decreasing with increasing density
Thermal coefficient	
of expansion	$10\text{–}12 \times 10^{-6}\ \text{K}^{-1}$
Thermal conductivity	0·55–0·60 W/mK for neat plasters
Strength	Transverse strength only useful for quality control
	Impact strength highest for D plasters
Hardness	Depending on density of set plaster
	Plasters containing lightweight aggregates are relatively soft
Fire-resistance	Improves fire resistance of building materials (depending on thickness of layer)
Corrosion of metals	Likely to cause corrosion of embedded metals, especially in damp conditions
Setting time	A usually 2–5 min., others 3–4 h. (expansion on setting)
Durability	Very durable when kept dry

[*Derived from the Construction Industry Handbook.*

water (which is a chemical process and followed by heat evolution) followed by the precipitation and crystallisation of calcium sulphate dihydrate from the supersaturated solution (which is a physical process).

Some important properties of gypsum building plasters are summarised in Table III.

TABLE IV: BS 1191 SPECIFICATIONS FOR GYPSUM
BUILDING PLASTERS

	A : Plaster of Paris	B: Retarded hemihydrate	C: Anhydrous gypsum	D: Keene's plaster
Chemical composition (% by weight of plaster)				
Min. SO_3	35%	35%	40%	47%
Min. CaO	$\frac{4}{5}SO_3$	$\frac{4}{5}SO_3$	$\frac{4}{5}SO_3$	$\frac{4}{5}SO_3$
Max. Na_2O + MgO	0·2%	0·2%	0·2%	0·2%
Ignition loss	4–9%	4–9%	max. 3%	max. 2%
Residue on BS14 sieve	max. 5%	max. 1%	max. 1%	max. 1%
Soundness	The set plaster pats shall show no signs of disintegration, popping or pitting when examined by the method described in 7 below			
Transverse strength (modulus of rupture)	min. 24·1 N/mm²	min. 11·7 N/mm²		
Mechanical resistance (dropping ball test) Diameter of indentation		max. 4·5 mm	max. 4·5 mm	max. 4·0 mm
(Linear) Expansion on setting		max. 0·2%		

BS TESTS

5. Chemical composition.

(a) *Determination of SO_3 and CaO content.* To a mix of finely-ground plaster (1 g) in water (100 ml) is added powdered ammonium carbonate (5 g). The whole is heated to boiling and kept simmering for 30 min. to remove any excess ammonium carbonate.

After neutralising the solution with 2N-*HCl* using a methyl red indicator, a further volume (20 ml) of *HCl* acid is added and the solution is kept boiling for a further 10 min. The insoluble material is filtered off and thoroughly washed with hot water. The filtrate and washings are combined and made up to 250 ml. The contents of SO_3 and CaO in the filtrate are then determined gravimetrically using standard procedures.

(b) *Determination of Na_2O and MgO.* To a mix of 1 g plaster in 100 ml water is added 1 g salt-free gypsum. The whole mixture is stirred occasionally for 1 h. and then allowed to stand until a clear supernatant liquid is obtained. The contents of soluble sodium salts and magnesium salts in the supernatant liquid are determined using standard procedures and expressed in terms of Na_2O and MgO.

6. Loss on ignition. By heating 2 g plaster to constant weight at a temperature 280–300 °C the loss in weight is determined and expressed as a percentage loss on ignition.

7. Soundness test. Six pats of neat plaster are prepared by mixing with an appropriate amount of water and filling in standard ring moulds (101·6 mm diameter and 6·35 mm deep). They are allowed to set in air of at least 80 per cent relative humidity for 16–24 h. (for A and B plasters) and for 3 days (for C and D plasters) before subjecting them to the action of saturated steam at atmospheric pressure for 3 h. without removing from the moulds. They are then examined visually for signs of disintegration, popping or pitting.

8. Determination of transverse strength. Six test prism specimens (101·6 mm long and 25·4 mm square in cross-section) are made from a standard six-compartmented metal mould. These prisms are allowed to stand for 24 h. under a damp cloth before being demoulded and dried to constant weight at 35–40 °C in a well-ventilated oven. They are tested dry in a transverse testing machine.

9. Mechanical resistance (dropping ball test). The dry prism test specimens are prepared as in **8** above. The mechanical resistance of the plaster is obtained by measuring the diameter of the indentation caused by dropping a hard steel ball (12·7 mm diameter and 8·33 g mass) from a height 1·82 m on to the horizontal surface of the specimen.

10. Determination of linear expansion on setting. About 200 g plaster are mixed with water to the standard final coat consistency and the paste is immediately filled into the cradle of the extensometer. The zero reading is adjusted and noted. The extensometer is then placed in a damp closet and left for 24 h. before the final reading is noted. The percentage linear expansion can then be calculated.

USES AND APPLICATIONS

11. Natural gypsum. Used:

    (*a*) As a set retarder in cement (3–5 per cent by weight).
    (*b*) In soil treatment.

(c) In chemical industries.

(d) As filler in paper, paint and pharmaceutical industries.

12. Calcined gypsum. Used in building industries:

(a) Wall and ceiling plaster (when mixed with sand, wood fibre, etc.).

(b) Pre-fabricated building products, such as wall boards, wall panels, floor tiles and bricks (when mixed with fibrous materials).

(c) Sound and heat insulation purposes (when mixed with lightweight aggregates such as perlite and vermiculite).

(d) In mortar (when mixed with slaked lime, cement and sand).

(e) Others including art and sculpture, dentistry.

Relevant British Standard Specifications

BS 1191:Part 1 Gypsum building plasters.
          Part 2 Premixed lightweight plasters.
CP211:Internal plastering.

PROGRESS TEST 2

1. Write down the chemical formula of gypsum. (p. 8)

2. Describe the changes which take place when gypsum is gradually heated from room temperature to about 1200 °C (p. 8)

3. Name the four classes of gypsum building plasters and compare their main properties and uses. (**1–4, 11–12**)

4. Describe, with the aid of chemical equations, the setting and hardening process of gypsum building plasters. (p. 11)

5. Outline the specifications and tests for gypsum building plasters. (Table IV, **5–10**)

EXAMINATION QUESTIONS

1. (a) What do you understand by the setting of a hydraulic material? Compare the setting processes of the following building materials:

(i) ordinary Portland cement,

(ii) building lime,

(iii) gypsum plaster.

(b) Discuss the characteristics and main uses of the following types of cement:

    (i) low heat Portland cement,
    (ii) sulphate resisting Portland cement,
    (iii) high alumina cement.

(S.O.E. Grad. Exam. C.E.)

2. Indicate the characteristics and main uses of the following types of plasters:

    (a) Class A type.
    (b) Class B type.
    (c) Class C type.
    (d) Class D type.

What are the precautions to be observed when decorating a newly-plastered surface?

(P.S.B. Pre-Inter. Arch.)

3. A survey of houses revealed the following maintenance defects:

    (i) staining on the internal wall of an external boiler chimney stack,
    (ii) popping of wall plaster,
    (iii) blistering of new paintwork,
    (iv) an old brick parapet wall out of plumb.

    (a) List TWO possible causes of each defect.
    (b) Select ONE cause for each defect, fully describe the reason for the defect and ways of remedying it.

(I.O.B. Final Part 1)

FURTHER READING

Materials and Technology, Vol. 2, Chapter 1, Longman/De Bussy, 1971.

CHAPTER III

BUILDING STONES

INTRODUCTION

Building stones which are derived from natural rocks have been used by man from the very beginning, first as tools, weapons and ornaments, later for building purposes.

The term "stone" usually signifies blocks or pieces of rock which have undergone cutting, sawing, shaping and sometimes polishing operations.

CLASSIFICATION

These natural rocks are classified according to their method of formation into three classes (*see* Table V):

(*a*) Igneous rocks.
(*b*) Sedimentary rocks.
(*c*) Metamorphic rocks.

1. Igneous rocks. These are rocks formed by cooling and consolidation of magma, the fluid melt of rock material (silicate slag) which developed on the earth several billion years ago.

Three categories of igneous rocks are recognised:

Plutonic rocks, when consolidated in large masses at great depths in the earth's crust. Such rocks are completely crystalline and coarsed-grained, due to the very slow cooling conditions, *e.g.* granite, syenite, diorite, gabbro, peridotite.

Hypabyssal rocks, when intruded in the form of small wall-like or sheet-like bodies, known respectively as dykes or sills near the surface of the earth's crust. Hypabyssal rocks are almost completely crystalline and finer-grained than plutonic rocks, due to faster cooling conditions, *e.g.* felsite, rhomb porphyry, dolerite, ilmenite.

Volcanic rocks, when extruded at the surface or under the sea as lava or produced from the product of volcanic explosions. Volcanic rocks are even finer-grained than hypabyssal rocks, due to the very fast cooling. They often contain some non-crystalline material such as glass, *e.g.* rhyolite, trachyte, andesite, basalt, obsidian, pitchstone, pumice.

TABLE V: CLASSIFICATION OF ROCKS

Igneous rocks: Plutonic	*Derived metamorphic rocks*
Granite⎫	
Syenite⎬	Gneiss
Diorite⎭	Amphibolites
Gabbro	Eclogites
Peridotite	Serpentine
Igneous rocks: Volcanic	
Rhyolite⎫	
Trachyte⎭	Sericite schists
Andesite⎫	
Basalt ⎭	Hornblende–Biotite–Schists
Sedimentary rocks	
Sandstone	Quartzite
Slates⎫	
Shale ⎭	Phyllites and mica schists
Limestone	Marble–Travertine

Igneous rocks can be divided, according to silica content, into four classes:

(i) *Acid:* containing more than 66 per cent silica, *e.g.* granites.

(ii) *Intermediate:* containing 52–66 per cent silica, *e.g.* syenite, diorite.

(iii) *Basic:* containing 45–52 per cent silica, *e.g.* basalts.

(iv) *Ultrabasic:* containing less than 45 per cent silica, *e.g.* ultrabasic appinite, quartz-dolerites.

2. Sedimentary rocks. These are deposits which were originally laid down as sediments under water or air. They can be *either* disintegration products of igneous rocks, transported by air or water, deposited, consolidated and cemented, *or* materials deposited from solution or through the agency of living organisms. Examples are gravels, sand clays, sandstones, shales, slates (obtained mechanically); peat, guano, limestone, chalk, coals, oil shales (obtained organically); borax, gypsum, anhydrite, rock salt (obtained chemically).

3. Metamorphic rocks. These are formed from igneous and sedimentary rocks by:

  (a) the action of heat and pressure, or
  (b) the action of chemical vapour, or
  (c) aqueous solution of mineral substances, and their deposition from solution, or
  (d) any combination of these.

Examples are gneiss, amphibolites, quartzite, phyllites, marble.

COMPOSITION OF ROCKS

The rocks from which building stones are derived usually contain silica minerals, silicate minerals or calcareous minerals.

4. Silica minerals. The main silica mineral is quartz (basically SiO_2), the chief constituent of sand. It is colourless, extremely durable and hard. It is insoluble in acids. Quartz is the main component of sandstone, quartzite or sand. Specific gravity 2·65, Moh's hardness 7·0.

5. Silicate minerals.

  (a) *Feldspar* which is composed of anhydrous aluminosilicates of potassium, sodium or calcium. It is less hard or durable than quartz. Colours may be red, pink or clear depending on composition. *Orthoclase* (potash) feldspar is basically of composition $K_2O.Al_2O_3.6SiO_2$. It is unaffected by acids. Specific gravity 2·57, Moh's hardness 6·0. Feldspar occurs in almost all igneous rocks, particularly in granite.

  (b) *Hornblende.* This is a silicate of Fe, Al, Mg, Ca, Na, K. It is strong and tough. Its hardness is the same as that of feldspar. Crystals can be green, brown or black, depending on composition. Specific gravity 3·0–3·4, Moh's hardness 5·0–6·0.

  (c) *Serpentine.* A silicate of Mg and Ca. Colours may be light green or yellow. It has no defined planes along which it splits readily. It is derived from partial decomposition of hornblende. Specific gravity 2·5–2·65, Moh's hardness 2·5–4·0.

  (d) *Mica.* Mainly a silicate of aluminium in combination

with iron, potassium or magnesium. It is a laminated material.

(i) *Muscovite* is silicate of *Al, K*. Specific gravity 2·76–3·0, Moh's hardness 2·0–2·5.

(ii) *Biotite* is silicate of *Al, Mg*. Specific gravity 2·7–3·1, Moh's hardness 2·5–3·0.

Mica occurs in various igneous (such as granite), sedimentary (such as clays, sands) and metmorphic rocks.

6. Calcareous minerals.

(a) *Calcite*, which is basically crystalline calcium carbonate ($CaCO_3$). Specific gravity 2·7, Moh's hardness 3·0. It occurs mainly in some sedimentary rocks (such as limestone, chalk and marble).

(b) *Dolomite* is a double carbonate of calcium and magnesium. It is frequently found in limestones. Specific gravity 2·9, Moh's hardness 3·5–4·0. Both calcite and dolomite are slightly soluble in pure water, but are readily attacked by mineral acids with the evolution of carbon dioxide.

COMMERCIAL BUILDING STONES

The following forms are available for construction purposes:

(a) Rubble stone (or fieldstone).
(b) Dimension stone (or cut stone).
(c) Flagstone (flat slabs).
(d) Crushed and broken stone.
(e) Stone dust or powder.

7. Rubble stone (or field stone). Stones of irregular shapes and sizes, which are either collected (fieldstones) or obtained from quarries, are made up into blocks or pieces for building walls, veneers, copings or sills.

8. Dimension or cut stone. Stones which are cut to specific size and thickness and finished at the stone mill. Cut stones are seldom used as structural members and are commonly used as masonry veneers, partitions, flooring, coping, sills, *etc.*

TABLE VI: PROPERTIES OF

Type of stone	Specific gravity	Compressive strength (N/mm^2)	Water absorption (%)	Moisture movement (%)	Thermal expansion ($\times 10^{-6}$ K^{-1})
GRANITE	2·4–2·9	90–146	0·1–0·5	0	3·0
SANDSTONE	2·4–3·0	21–105	2·0–8·5	0–0·07	ca 12
SLATE	2·4–2·9	75–200	0·1	negligible	ca 11
LIMESTONE	2·1–2·4	9–59	2·5–11·0	ca 0·01	ca 4
MARBLE	2·7–2·9	ca 60	0·1–0·5	negligible	ca 4
QUARTZITE	ca 2·6	ca 100	0·1–0·5	negligible	ca 11

9. Flagstone (flat slabs). Flat slabs of thin stone (25–50 mm thick) are used for paths, walks, terraces, flooring, coping, sills, roofing, *etc.*

10. Crushed and broken stone. Stones crushed or broken mechanically vary in shapes and sizes and have usually rough surface texture and sharp edges. They are used as aggregates for concrete work, roadwork, etc.

11. Stone dust or powder. Used for surfacing asphalt paving, as filler in paints, resilient flooring, etc.

COMMON BUILDING STONES

Thermal conductivity (W/mK)	Frost-resistance	Acid-resistance	Composition	Uses
3·0	good to excellent	good	mainly feldspar quartz and mica	walling, cladding, plinths, window surrounds and steps
ca 1·3	poor to excellent	good (except calcareous types)	mainly quartz with some mica, feldspar	paving, walling, cladding and coping
ca 1·9	good to excellent	good	mainly SiO_2 Al_2O_3 and iron oxides	cladding, sills, coping, steps, paving, roofing
ca 1·5	poor to very good	poor	mainly $CaCO_3$	walling and cladding
ca 2·5	good to excellent	poor	mainly $CaCO_3$	window surrounds, cladding, floors, stairs
ca 3·0	good to excellent	good	mainly quartz	cladding, plinths, pavings, floors, stairs

[*Mainly derived from the Constuction Industry Handbook.*

PROPERTIES OF BUILDING STONES

The main requirements of building stones are:

12. Strength. Many types of stones are generally strong for construction purposes (a compressive strength of 35 N/mm^2 is considered to be satisfactory) with the exception of some weaker forms of limestones and sandstones.

13. Durability. Stones are generally extremely durable and withstand well the adverse effects of weathering, heat and aggressive chemicals. Durability is dependent on porosity and on composition.

14. Hardness. Very variable—some soft sandstones can be scratched easily while some granites can be harder than steel.

15. Ease of working. This affects cost in producing stones of required sizes and shapes, and is related to hardness of the material.

16. Appearance. For aesthetic purposes—colour, grain and texture of stones.

17. Availability. This, too, affects cost, as transportation over long distance can be cumbersome and expensive.

18. Maintenance. Ideally, stones should be easy and cheap to maintain and clean. This proves to be a problem in practice for many building stones.

Important properties of some common building stones are summarised in Table VI.

Relevant Specifications and Codes of practice:

CP 121:Part I:1973	Walling. Brick and block masonry
CP 121.201:1951	Masonry walls ashlared with natural stone or with cast stone.
CP 121.202:1951	Masonry. Rubble walls.
CP 298:1972	Natural stone cladding (non-loadbearing).
BS 435:1975	Dressed natural stone kerbs, channels, quadrants and setts.
BS 1217:1975	Cast stone.
BS 2847:1975	Glossary of terms for stone used in building.

PROGRESS TEST 3

1. How may natural rocks be classified? (p. 16)
2. Describe, giving examples, the various classes of natural rocks. (1–3)
3. What are the compositions of building stones? (4–6)

4. Outline the various forms of building stones available commercially. **(7–11)**

5. Summarise the main properties of the different types of building stones. **(12–18)**

EXAMINATION QUESTION

1. Discuss the difficulties existing in assessing the frost resistance of building stones by means of laboratory tests. (I.O.B. Assoc. Part 1)

FURTHER READING

O'Neill, H., *Stone for building*, British Stone Federation, 1965.
Schaffer, R. J., *The weathering of natural building stones*, D.S.I.R. Building Research Special report No. 18, reprinted 1972.

CHAPTER IV

CEMENT AND CONCRETE

INTRODUCTION

The term "cement" used in the construction field applies to materials of calcareous origin for building together fragments of stones, sand, bricks, etc.

"Mortar" implies a mixture of cement and sand, whereas "concrete" implies a mixture of cement, sand and stones. Water, of course, must be added in order to make mortar or concrete.

PORTLAND CEMENT

The term "Portland" probably owes its name to its resemblance to Portland Stone (a limestone quarried in Dorset) after this cement has set into a hard mass. The invention of this cement in the United Kingdom must be attributed, at least partially, to John Smeaton (1756) who discovered the hydraulicity of lime, to Joseph Aspdin (1824) who first patented Portland cement and to Isaac C. Johnson (1845) who improved on the process of manufacture.

1. Manufacture. Manufacture involves two distinct stages: the mixing of the raw materials followed by the heating process whereby chemical action takes place (*see* Fig. 1).

Raw materials. The raw materials used are chalk or limestone (both basically calcium carbonate) and clay or shale (basically alumino-silicates). The raw materials are mixed in appropriate proportions and finely ground either in water ("wet" process) or in a dry condition ("dry" process) depending on the nature of raw materials used.

The heating process. The raw materials, either dry or as a slurry, are fed in at the upper end of a large rotary kiln which is rotating slowly about its axis inclined a few degrees from the horizontal. Heat is applied from the lower end of the kiln by injecting fuel (pulverised coal, oil or natural gas) and air. As the mixture works its way along the kiln, it is subjected to a gradual rise in temperature (up to 1500 °C) and undergoes successive chemical reactions. Sintering occurs at approximately 1300–1400 °C and the material

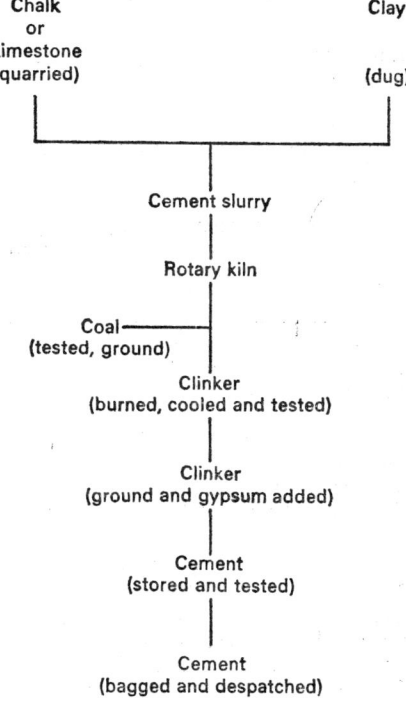

Fig. 1.—*Manufacture of Portland cement.*

fuses into small balls known as cement clinker. After cooling, the cement clinker is mixed with a small amount of gypsum and finely ground. The purpose of adding gypsum is to control the rate of setting of the cement.

2. Chemical composition. The successive thermal reactions taking place inside the kiln are:

(a) Loss of water from the mixture of raw materials (dehydration).

(b) Loss of carbon dioxide from calcium carbonate in chalk or limestone (decarbonation) leaving calcium oxide.

(c) Fusion of the oxides: calcium oxide (from chalk or limestone), silica, alumina and ferric oxide (from clay or shale).

(d) Chemical combination of these oxides followed by crystallisation on cooling.

Four main components (mineralogical compounds) are basically present in the cement clinker and these are listed in Table VII.

TABLE VII: PORTLAND CEMENT

Component	Chemical formula	Abbreviated notation
Tricalcium silicate	$3\,CaO\,.\,SiO_2$	C_3S
Dicalcium silicate	$2\,CaO\,.\,SiO_2$	C_2S
Tricalcium aluminate	$3\,CaO\,.\,Al_2O_3$	C_3A
Tetracalcium aluminoferrite	$4\,CaO\,.\,Al_2O_3\,.\,Fe_2O_3$	C_4AF

NOTE: For simplicity cement chemists preferred to use the shortened notation by representing each oxide by one letter symbol:

$CaO = C$; $\qquad\qquad$ $Fe_2O_3 = F$;
$SiO_2 = S$; $\qquad\qquad$ $H_2O = H$;
$Al_2O_3 = A$; $\qquad\qquad$ $Ca(OH)_2 = CaO.H_2O = CH$.

Minor oxides, such as magnesia (MgO), titanium oxide (TiO_2), manganese oxide (Mn_2O_3), the oxides of alkalis (K_2O and Na_2O), are usually present in much smaller amounts.

The properties of cement and concrete depend partly on the presence of these compounds (*see* Table VIII).

C_3S is the principal active hydraulic compound of the cement and the early strength of cement is dependent almost entirely on C_3S. C_2S is slowly hydraulic especially in the β-form, but catches up with C_3S in strength and hydraulicity eventually (*see* Fig. 2).

The hydration products of C_3S and C_2S are the tobermorite gel and calcium hydroxide. The fineness of the tobermorite particles, having a specific surface of the order of 300 m^2/g, is responsible for the cementing properties as well as other important engineering properties (such as strength, shrinkage,

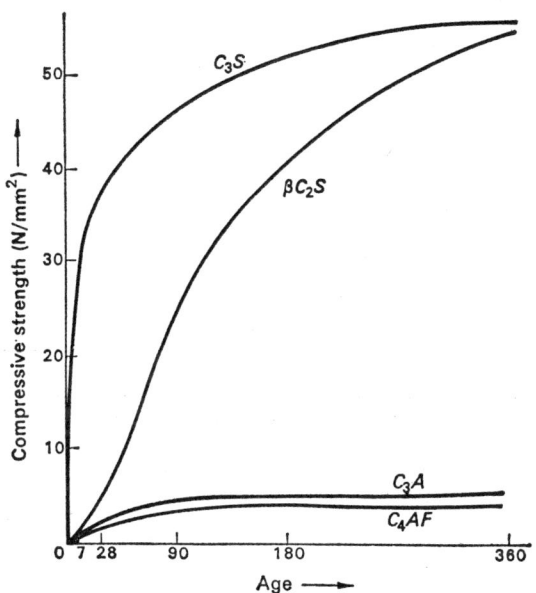

Courtesy Bogue and Lerch

FIG. 2.—*Compressive strengths of pastes made from C_3S, βC_2S, C_3A and C_4AF.*

permeability and resistance to the stresses of freezing and thawing). Calcium hydroxide provides the alkaline medium which is beneficial for the protection against corrosion of reinforcing steels.

C_3A is rapidly hydrated with water with evolution of considerable heat, but the product has little cementing value.

C_3A in cement clinker is responsible for the flash setting of the cement on addition of water. In order to control this fast hydration, calculated amounts of gypsum ($CaSO_4.2H_2O$) of the

order of 1–5 per cent by weight of cement are added. This leads
to the formation of an expansive compound—calcium sulpho-
aluminate hydrate. The durability of concrete in sulphate
water is therefore governed by the C_3A content. When a
hardened concrete is attacked by sulphate ions, the crystals
form in the pores of the structure with a volume increase of
about 227 per cent, which may bring about a disintegration
of the structure. Therefore, a cement containing a smaller
C_3A content will be less prone to sulphate attack.

C_4AF has also no cementing value. It hydrates in the same
way as C_3A. Magnesia is present in Portland cement as MgO
or periclase, which is limited in the specifications, because it
may give rise to excessive expansion.

A small amount of lime also remains uncombined in the
clinker as free CaO, which may give rise to excessive expansion
(known as "unsoundness").

TABLE VIII: INFLUENCE OF MINERALOGICAL COMPOUNDS OF
CEMENT

Compound	Effect on cement
C_3S and C_2S	Strength characteristics of Portland cement, C_3S develops higher early strength and higher evolution of heat than C_2S
C_3A	Very rapid hydration Reacts with sulphates to form an expansive compound, responsible for the so-called "sulphate attack"
C_4AF	Similar to C_3A Presence of Fe_2O_3 is responsible for the grey colour of Portland cement

The "potential" compound composition of Portland cement
can be approximately evaluated from chemical analysis data
using *Bogue's equations*, which assume the absence of minor
compounds.

Percentage values of mineralogical compounds are given by:

$$C_3S = 4\cdot07\ (CaO) - 7\cdot60\ (SiO_2) - 6\cdot72\ (Al_2O_3) - 1\cdot43\ (Fe_2O_3) - 2\cdot85\ (SO_3)$$
$$C_2S = 2\cdot87\ (SiO_2) - 0\cdot754\ (C_3S)$$
$$C_3A = 2\cdot65\ (Al_2O_3) - 1\cdot69\ (Fe_2O_3)$$
$$C_4AF = 3\cdot04\ (Fe_2O_3)$$

where (   ) represents the percentage of the given oxide or compound.

3. Hydration. The term "hydration" is used to include all reactions of cement with water, that is to say, to both true hydration and hydrolysis. The cement components react with water as follows (abbreviated notations are used):

C_3S: $2C_3S + 6H \rightarrow C_3S_2H_3 + 3CH$
                                      tobermorite   calcium
                                           gel   hydroxide

C_2S: $2C_2S + 4H \rightarrow C_3S_2H_3 + CH$

C_3A: $C_3A + 6H \rightarrow C_3AH_6$
                                           calcium
                                         aluminate
                                         hydrate
                                         (stable)

                   C_3A + gypsum + water $\longrightarrow$
                             calcium sulphoaluminate
                             $(3CaO.Al_2O_3.3CaSO_4.31H_2O)$
                             "Ettringite" (an expansive compound)

$C_4AF(C_6A_2F - C_6AF_2)$: $C_4AF + 2CH + 10H \longrightarrow$
                                       $C_3AH_6 + C_3FH_6$
                                         ferrite
                                         hydrate

                   C_4AF + gypsum + water $\longrightarrow$
                             calcium sulphoaluminate +
                                     calcium sulphoferrite

The rate of reaction and the heat of hydration for these components are given in Table IX.

TABLE IX: HYDRATION OF CEMENT COMPONENTS

Type of reaction	Heat of hydration (kJ/kg)	Rate of reaction (80% hydration) Time (days)
C_3S + H_2O	504	10
C_2S + H_2O	252	100
C_3A + H_2O	869	—
C_3A + gypsum + H_2O	1365	—
C_3A + H_2O + $Ca(OH)_2$	840	6
C_4AF + H_2O + $Ca(OH)_2$	420	50

4. Setting and hardening. When water is added to cement to form a cement paste, this paste gradually changes with time from a fluid to a rigid state. The term "setting" is used to describe this stiffening state. Once set, the cement paste gradually develops strength and forms a hard mass. This is referred to as "hardening".

The process of setting and hardening is merely one of hydration of cement components. There has been much controversy

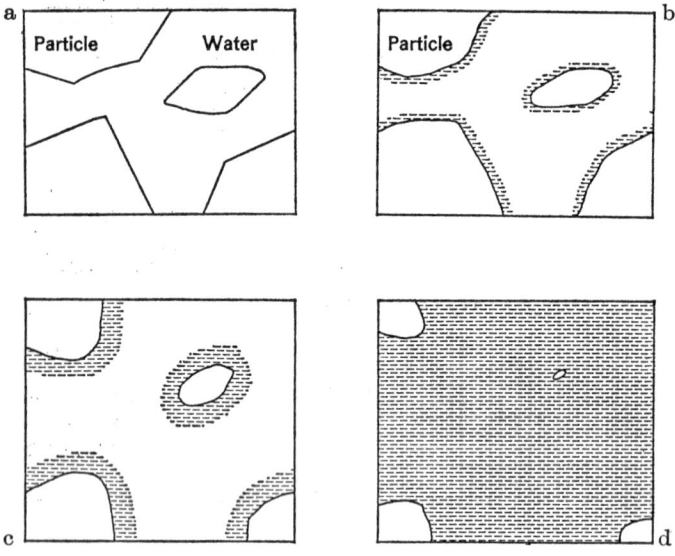

FIG. 3.—*Setting and hardening of cement.*

(a) (*Top left*) *Dispersion of unreacted cement particles in water.*
(b) (*Top right*) *Hydration products appear after a few minutes.*
(c) (*Bottom left*) *More hydration products: setting takes place after a few hours.*
(d) (*Bottom right*) *Formation of gel: hardening takes place.*

as regards the precise mechanism of setting and hardening. In the early days of cement chemistry two main hypotheses were put forward: Le Chatelier's Crystallisation Hypothesis and Michaëlis' Gel Hypothesis.

It is believed that setting and hardening take place in four stages, as shown schematically in Fig. 3.

5. Tests on cement. Owing to the complex nature of Portland cement, stringent control is necessary to ensure the uniformity and consistency of the cement produced.

(a) Chemical tests (BSS 12, 4550). Chemical tests are based on the gravimetric determination of the following oxides: CaO, SiO_2, Al_2O_3, MgO, K_2O, Na_2O. The cement sample is digested in an acid solution, followed by selective precipitation and ignition into the oxides.

Faster analytical methods are now available—they are based on complexometric and spectrophotometric techniques (BS 4550: Part 2: 1970 Chemical tests).

(b) Physical tests.

(i) *Fineness.* Fineness is measured in terms of specific surface, which is the total surface area of all the particles per unit mass of cement (m^2/kg).

The method described in the British Standards is the air-permeability method developed by Lea and Nurse and it is based on the relationship between the flow of air through a layer of cement and the specific surface of the particles (BS 12).

(ii) *Strength* (BS 12). Two standard methods exist of testing the compressive strength of cement: the cement mortar test and the concrete test.

In the mortar test, a 1:3 cement/sand mortar with a water/cement ratio of 0·4 is used. 70·7 mm (2·78 in) cubes are made, each cube requiring a mix proportion of 185 g cement, 555 g Leighton Buzzard Standard sand and 74 g (74 ml) water. They are compacted on a standard vibrating table for two minutes at a frequency of 200 hertz (vibrations per second). The cubes are kept moist for 24 hr. before demoulding and immersed in water at room temperature until tested in a wet condition.

In the concrete test, 101 mm (4 in) cube moulds are used. Each cube is made by mixing 325 g of cement, 195 g of water (corresponding to a water/cement ratio of 0·60) and an appropriate amount of fine and coarse aggregates (*i.e.* sand and stone) so as to give a workable concrete with a slump of 12·7–50·8 mm ($\frac{1}{2}$–2 in.). The cubes are cured and tested in a wet condition as in the mortar test.

The tensile strength for Portland cement is of lesser importance—it is done by testing briquettes (made from 1:3 cement/

sand mortar and water/cement ratio of 0·32) in direct tension.

(*iii*) *Setting time.* Two stages of setting are arbitrarily chosen—the initial set (corresponding to a rapid rise in temperature when cement and water are mixed) and the final set (corresponding to the peak temperature obtained).

The setting times of the neat cement paste are measured by the Vicat apparatus by using a needle (1 mm²) for the initial setting time and by using a needle (1 mm²) fitted with a special attachment for the final setting time. See BS 12 for details of the procedures.

(*iv*) *Soundness.* The soundness test is designed to detect in Portland cement the presence of any appreciable excess of free lime which would cause expansion after subsequent hydration in hardened cement paste. Magnesia (*MgO*), like free lime, will also cause unsoundness of cement. Soundness is determined by using the Le Chatelier method, measuring the expansion (in mm) between two pointers in the mould containing the paste when subjected to boiling in water (accelerated hydration). BS 12 gives details of this test.

(*v*) *Heat of hydration* (for low-heat cement). This is the energy (heat) in J/kg, evolved upon complete hydration of cement at a given temperature. The calorimetric method described in BS 1370 is mostly used to test low-heat Portland cement. Table XV gives the heats of hydration of different types of cement.

TYPES OF CEMENT

6. Portland cement types.

(*a*) *Ordinary Portland Cement* (*OPC*). This is the most-used cement and is suitable for most types of work. Its properties are covered by BS 12 (*see* Tables X and XI). It has a medium rate of setting and hardening.

(*b*) *Rapid-hardening Portland cement* (*RHPC*). This has a composition similar to that of OPC, the main difference being its finer grinding and possibly higher C_3S content. Consequently, it has a high rate of hardening and strength development, although its setting times are similar. It is also covered by BS 12 (*see* Table X for composition and Table XI for properties). It is useful in situations where early stripping of formwork is required and in cold-weather concreting (due to its higher rate of heat evolution).

(*c*) *Extra Rapid-hardening Portland cement* (*ERHPC*). This is obtained by grinding RHPC with about 2 per cent calcium chloride (commercial grade) or by much finer grinding of RHPC with or without the addition of $CaCl_2$, which acts as a setting and hardening accelerator. The effect of corrosion of reinforcing steel may be questionable but it may be serious in the case of prestressing steel wires.

The strength obtained is about 25 per cent higher than that of RHPC at 1 or 2 days and 10–20 per cent higher at 7

TABLE X: TYPICAL COMPOSITION OF PORTLAND CEMENTS *

											Igni-
						Na_2O					tion
						+					loss,
Cement	CaO	SiO_2	Al_2O_3	Fe_2O_3	MgO	K_2O	SO_3	S^{2-}	TiO_2	Mn_2O_3	etc.
OPC	64·78	21·31	6·06	2·77	0·86	0·91	1·91	—	0·33	0·11	2·27
RHPC	64·30	20·82	5·25	3·01	1·13	0·96	2·01	—	0·35	0·07	0·94
LHPC	61·90	25·44	4·57	2·07	1·64	0·90	2·22	—	0·24	0·08	2·10
SRPC	62·40	21·15	4·03	5·63	0·88	0·88	2·36	—	0·28	0·12	0·96

[Figgs, J. W. and Bowden, C. R., *Analysis of concrete, BRS publications, H.M.S.O., 1971.*
* British types

days. It is useful for concreting in cold frosty weather but should not be used in prestressed concrete.

(d) *Low-heat Portland cement (LHPC)*. Its composition is similar to that of OPC or RHPC but is adjusted to reduce the heat of hydration. This is achieved by having a reduction in C_3S and a proportional increase in C_2S. It has a comparatively slower rate of strength development, although its ultimate strength is the same as that of OPC or RHPC. It is also more resistant to sulphate attack owing to the reduction in C_3A (*see* Tables X and XI). It is suitable for use in massive structures such as dams, retaining walls, bridge abutments, etc. where the rise in temperature due to heat developed by the hydration of cement can cause serious cracking of the structure and where high early strength is not required. It is covered by BS 1370, which limits its heat of hydration to not more than 251 J/g at 7 days and 293 J/g at 28 days.

(e) *Portland blastfurnace cement (PBFC)*. This is obtained by grinding Portland cement clinker with 30–75 per cent blastfurnace slag (BS 146 prescribes an amount not exceeding 65 per cent by weight of the slag). Blastfurnace slag, a waste product in the manufacture of pig iron, has the same components as Portland cement except for a lower *CaO* and a

TABLE XI: PROPERTIES OF VARIOUS TYPES OF CEMENT

Type	OPC (BS 12)	RHPC (BS 12)	LHPC (BS 1370)	PBFC (BS 146)
Composition	$CaO + SiO_2$, Al_2O_3, Fe_2O_3	$CaO + SiO_2$, Al_2O_3, Fe_2O_3	$CaO + SiO_2$, Al_2O_3, Fe_2O_3	Portland Cement clinker + Granulated blastfurnace slag (max. 65%)
Fineness Min. specific surface	225 m²/kg	325 m²/kg	320 m²/kg	225 m²/kg
Chemical composition				
I.R.[1] (max.)	1·5%			1·5%
MgO (max.)	4·0%			7·0%
SO₃ (max.)	2·5% (when C₃A 7% or less) 3·0% (when C₃A 7%)			3·0%
I.L.[1] (max.)				1·5%
temperate climate:	Max. 3·0%			Max 3·0%
tropical climate:	Max. 4·0%			Max. 4·0%
Compressive strength (min.)	Mortar: (1:3)			
3 days:	15·1 N/mm²	20·7 N/mm²	7·5 N/mm²	15·1 N/mm²
7 days:	23·4 N/mm²	27·6 N/mm²	13·8 N/mm²	23·4 N/mm² 28 days: 34·4 N/mm²
	Concrete: (1:2:4)			
3 days:	8·3 N/mm²	11·7 N/mm²	3·4 N/mm²	8·3 N/mm²
7 days:	13·8 N/mm²	17·2 N/mm²	13·8 N/mm²	13·8 N/mm² 28 days: 22·0 N/mm²
Setting time Initial	Min. 45'	Min 45'	Min 1 hr.	Min 45'
Final	Max. 10 hr.	Max. 10 hr.	Max. 10 hr.	Max. 10 hr.
Soundness	Max. 10 mm; Retest 5 mm	Max. 10 mm	Max. 10 mm 251 J/g (7 days) 293 J/g (28 days)	Max. 10 mm (5 min.)
Heat of hydration				
Tensile strength		Min. 2·1 N/mm² at 24 hr.		

Type	SRC (BS 4027)	SSC (BS 4249)	LHPBFC (BS 4246)	HAC (BS 915)*
Composition	$CaO + SiO_2$, Al_2O_3, Fe_2O_3	Granulated blastfurance slag (min. 75%) + $CaSO_4$ + Portland Cement clinker (or lime)	Granulated blastfurance slag (50–90%) + Portland cement clinker	$CaO + Al_2O_3$
Fineness Min. specific surface	250 m²/kg	400 m²/kg	275 m²/kg	225 m²/kg
Chemical composition I.R.¹ (max.)	1·5%	3·0%	1·5%	Al_2O_3 Min. 32%
MgO (max.)	4·0%	8·0%	9·0%	Al_2O_3/CaO ratio 0·85–1·3
SO₃ (max.)	2·5%	4·5%	3·0%	
I.L.¹ S¹ temperate climate:	Max. 3·0%	1·5%	2·0%	
tropical climate	Max. 4·0%			
Compressive strength (min.)			7·6 N/mm²	1 day: 41·3 N/mm²; 48·2 N/mm²
3 days: (*Mortar*)	15·1 N/mm²	13·8 N/mm²		
7 days:	23·4 N/mm²	23·4 N/mm²		
(*Concrete*)		34·4 N/mm²		
3 days:	8·3 N/mm²	6·9 N/mm²		
7 days:	13·8 N/mm²	16·5 N/mm²		
		25·5 N/mm²		
Setting time Initial	Min. 45′	Min. 45′	Min. 1 hr.	Min. 2 hr.
Final	Max. 10 hr.	Max. 10 hr.	Max. 15 hr.	Max. 6 hr.
Soundness	Max. 10 mm	5 mm (modified)	Max. 10 mm; Retest Max. 5 mm	Max. 1 mm
Heat of hydration		Max. 251 J/g (7 days) Max. 293 J/g (28 days)	Max. 251 J/g (7 days) Max. 293 J/g (28 days)	
Tensile strength				

¹ I.R. = Insoluble Residue. I.L. = Ignition Loss.

*BS 915 is now withdrawn

higher SiO_2 content. PBFC, therefore, has a lower heat of hydration, a slower rate of setting and hardening than those of OPC and RHPC, although its ultimate strength is unaffected. It is fairly highly resistant to sulphate attack, due to its low C_3A content. It is suitable for use in reinforced concrete, water-retaining structures, precast objects such as concrete pipes, mass concrete and sea water construction. It is covered by BS 146 (see Tables XI and XII).

(f) *Low-heat Portland blastfurnace cement (LHPBFC)*. This consists of an intimate mixture of Portland cement

TABLE XII: TYPICAL COMPOSITION OF PBFC, SSC AND HAC*

	Typical composition (%)										
Cement	CaO	SiO_2	Al_2O_3	Fe_2O_3	FeO	Na_2O $+$ K_2O	SO_3	S^{2-}	TiO_2	Mn_2O_3 MgO	Loss on ignition, etc.
PBFC	56·58	24·38	9·49	2·13	trace	0·84	1·55	0·30	0·30	0·53	2·51	1·39
SSC	45·41	25·77	12·76	0·86	trace	1·27	7·53	0·99	0·54	1·21	2·80	0·86
HAC	38·88	5·27	38·33	9·17	5·86	0·34	0·13	—	1·83	0·08	0·42	0·19

[*Figg J. W. and Bowden C. R., Analysis of concrete, BRS publications, H.M.S.O., 1971.*
* British types

clinker and 50–90 per cent granulated blastfurnace slag. It is covered by BS 4246 whose requirements are summarised in Table XI.

(g) *Sulphate-resisting Portland cement (SRPC)*. As its name implies, this cement is characterised by its high sulphate-resistance, so that it is recommended for use in subsoils containing up to 2 per cent SO_3 and in ground waters containing up to 5000 mg SO_3/litre solution. This sulphate-resistance is achieved by having a low C_3A content in its composition It is covered by BS 4027 which limits its C_3A content to not more than 3·5 per cent when calculated by using Bogue's equations. In many respects, SRPC is similar to OPC (see Tables X and XI).

Sulphate-resistance can be tested by immersing the cement specimens in sulphate solutions followed by observations, measurements and/or mechanical testing. This is very tedious and time-consuming. Accelerated testing by using concentrated sulphate solutions, however, is far from satisfactory and correlation may prove difficult or inconclusive.

(h) *White and Coloured Portland cements*. The grey colour of OPC is due to the presence of an appreciable amount of iron oxides. To obtain *white Portland cement*, this must be reduced. This is achieved by proper selection of the raw materials (using the pure form of clay such as China clay) and the fuel used in the firing process (*e.g.* using oil in place of coal). White Portland cement is more expensive but is not usually as strong as OPC. Nevertheless, it readily satisfies the requirements of BS 12.

A typical compound composition of white Portland cement is given in Table XIII.

TABLE XIII: TYPICAL COMPOUND COMPOSITION
OF WHITE PORTLAND CEMENT

Compound	Content (%)
C_3S	51·0
C_2S	26·0
C_3A	11·0
C_4AF	1·0
SO_3	2·6
alkalis ($Na_2O + K_2O$)	0·25

[A. M. Neville, Properties of Concrete, Pitman,
second edition, 1973.

"Snowcrete" is a white Portland cement produced commercially in the U.K.

Coloured Portland cements can be obtained by grinding white Portland cement (or sometimes OPC) with 5–10 per cent pigment. The pigment added should be chemically inert so that it does not interfere with the cement hydration and it should conform to BS 1014: Pigments for cement magnesium oxychloride and concrete. Some pigments used are given in Table XIV.

TABLE XIV: PIGMENTS USED IN CEMENT

Pigment	Colour of product
Carbon black	black or dark
Chromium oxide	green
Ultramarine blue	blue
Manganese dioxide	black or brown
Iron oxides	yellow, red, brown or black
Cobalt blue	blue

7. Other types of cement.

(a) *High-alumina cement* (*HAC*). This is obtained by heating a mixture of limestone or chalk and bauxite (an aluminium ore) to fusion temperatures (1500–1600 °C) and grinding the resulting clinker. Its composition is very different from that of Portland cement, containing a much higher Al_2O_3 content but a lower CaO and SiO_2 content (*see* Table XII). It is covered by BS 915, which requires:

$$\text{Minimum } Al_2O_3 = 32\%$$
$$\text{Ratio } \frac{Al_2O_3}{CaO} = 0.85 \rightarrow 1.3$$

Table XI summarises its requirements according to BS 915. Its compound composition is much less known than that of OPC. The following results are obtained microscopically by Robson.

CA	60–65%
Iron compounds	high
C_2S	10%
glass	5%
C_2AS	2%
$C_{12}A_7$	low

In addition, $C_6A_4 . FeO . S$ and $C_6A_4 . MgO . S$ may be also present.

(*i*) *Hydration.* The hydration process of the main clinker minerals can be represented as follows:

$$2CA + 11H \xrightarrow{\text{at normal temp.}} C_2AH_8 + AH_3 \quad \text{. (1)}$$

$$C_{12}A_7 + 11H \xrightarrow[\text{water}]{\text{at normal temp.}} 6C_2AH_8 + \underset{\substack{\text{aluminium} \\ \text{hydroxide}}}{AH_3} \quad \text{. (2)}$$

At lower temperatures, the hydration product CAH_{10} may be formed predominantly.

At higher temperatures and/or longer duration of time, the following conversion may take place:

$$3C_2AH_8 \rightarrow 2C_3AH_6 + AH_3 + 9H. \quad \text{. (3)}$$
$$\underset{\substack{\text{hexagonal} \\ \text{crystal system}}}{} \quad \underset{\substack{\text{cubic crystal} \\ \text{system}}}{}$$

Consequently, there follows a considerable reduction in volume and at conventional W/C ratio of 0.5–0.6, this may result in increased porosity and permeability and a drastic reduction in strength. The water required for hydration of high-alumina cement is calculated to be about 55 per cent, which is about twice the water of hydration of OPC. However, the tendency for the conversion (*see* equation (3)) to take place makes it possible and highly desirable to use a lower W/C in order to obtain higher strengths and permeabilities and consequently lower porosities.

The rate of hydration (and hence the rate of strength development) is relatively higher than that of OPC and RHPC, although the total heat of hydration is similar for all of them.

(*ii*) *Resistance to chemical attack.* The reasons for the chemical resistance of HAC are:

    I. Absence of lime $(Ca(OH)_2)$.
    II. Presence of alumina gel $(Al(OH)_3)$.
    III. Chemical stability of the aluminates.

It is extremely resistant to sulphate solutions. It is also resistant to soft water, moor water, sea water, dilute hydrochloric acid and weak acids (*e.g.* carbonic acids). But it is attacked by alkalis which dissolve the alumina gel to form alkali aluminate. In the presence of high temperature and humidity conversion of HAC concrete takes place, resulting in loss of strength and deterioration.

(*iii*) *Uses.* HAC is about three times as expensive as OPC. It is used to resist attack by sulphates (in subsoils and ground waters) and where very high early strength is required. Because of the amount of heat of hydration evolved, care should be taken not to allow its temperature to rise above 29 °C as this would result in serious loss of strength. It is also used as a refractory concrete when mixed with a refractory aggregate such as crushed firebrick. Such concrete is stable up to a temperature of about 1300 °C. By using special refractory aggregates, such as fused alumina or carborundum, it can be used for higher temperatures (up to 1600 °C).

Ordinary Portland cement should not, as a rule, be mixed with high-alumina cement because this would cause a flash set, although mixtures of the two cements in judicious proportions have been used when an accelerated setting is required (*e.g.* for stopping the ingress of water).

(*b*) *Supersulphated cement (SSC).* This is obtained from an intimate mixture of 80–85 per cent granulated blastfurnace slag, 10–15 per cent calcium sulphate (anhydrite or dead-burnt gypsum) and about 5 per cent Portland cement clinker.

A typical composition is given in Table XII, and its requirements according to BS 4248 are summarised in Table XI.

The characteristic properties of SSC are:

(*i*) Extreme fineness: its specific surface is 400–500 m²/kg, which is well within the minimum of 400 m²/kg specified by the BS.

(*ii*) Low heat hydration during setting (due to comparatively low CaO content): 170–190 J/g at 7 days, 190–210 J/g at 28 days (*see* Table XV for comparison with other types of cement). This property makes it very suitable for mass concrete work, *e.g.* dams.

(*iii*) Low drying shrinkage of 0·2 mm/m, as compared with the value of 1·0 mm/m for OPC.

(*iv*) Good resistance to chemicals such as:

    I. Calcium and sodium sulphates (except magnesium sulphate and ammonium sulphate at a concentration of 350 parts SO_3 per 100 000).

    II. Chlorides, alkali hydroxides and carbonates.

    III. Linseed and other vegetable oils.

    IV. Sugars, phenols and cresols.

    V. Weak solutions of mineral acids down to pH 3·5.

TABLE XV: TOTAL HEATS OF HYDRATION OF CEMENTS

Cement	Heat of hydration (J/g) at ages of:					
	1 day	2 days	3 days	7 days	28 days	90 days
OPC	96–192	121–221	175–271	196–314	276–393	335–439
RHPC	146–296		188–372	213–380	293–418	
LHPC			188	230	271	314
PBFC	84–109	117–196	126–260	167–293	293–355	314–377
SSC				167–188	188–209[1]	
HAC	322–390	326–393	330–397			

[*D. F. Orchard, Concrete Technology Vol. 1, Applied Science Publishers Ltd., London, 3rd edition, 1973.*

[1] F. M. Lea, *The Chemistry of cement and concrete*, Edward Arnold Ltd., 3rd edition, 1970.

SSC should be stored dry, free from moisture and atmospheric carbon dioxide. It requires more water for full hydration (W/C 0·6–0·7) than OPC does. It is used in sulphate-containing subsoils and ground waters, in mass concrete such as in the building of dams and in sea water construction.

(c) *Masonry cement.* This cement is used, as its name implies, in masonry mortars for bonding bricks, blocks, tiles, etc. with the purpose of improving plasticity and water-retentivity and reducing shrinkage.

It is obtained by grinding an intimate mixture of very finely-ground Portland cement, an inert filler such as limestone and an air-entraining agent such as calcium stearate or some other plasticisers.

Because of the air-entrainment, the strength of the masonry cement is usually lower than that of OPC and is ideally suited for brick construction. However, it is not suitable for use in structural concrete. In general, a 1:3·5 masonry cement:sand mix is equivalent to a 1:1:6 OPC:lime:sand mix and a 1:5 masonry cement:sand mix is equivalent to a 1:3:12 OPC:lime:sand mix.

As yet, it is not covered by any BS, but the American counterpart, ASTM C 91–71, stipulates the following requirements for masonry cement:

Fineness:	Residue on a No. 325 sieve (44μm) = 24 per cent (maximum)
Chemical composition:	CaO = 69–72 per cent
	SiO_2, Al_2O_3, Fe_2O_3 and MgO = 1 per cent (maximum)
	Ignition loss = 23–27 per cent
Compressive strength:	For a 1:3 mortar mix:
	Minimum 3·4 N/mm² (at 7 days)
	Minimum 6·2 N/mm² (at 28 days)
Setting time	Initial: Minimum 2 hr.
(Gillmore method):	Final: Maximum 24 hr.
Soundness:	Autoclave expansion: maximum 1·0 per cent
Air content of mortar:	12–22 per cent (by volume)
Water retention:	flow after suction for 60 sec., when tested on a flow table, to be not less than 70 per cent of original flow.

(d) *Expanding cements* (or *Expansive cements*). These are cements which are apparently non-shrinking or which expand during the hardening process. One such type of expanding cement is obtained by an intimate mixture of Portland cement, an expanding agent and a stabiliser.

Cementing material: Portland cement (approximately 70 per cent)

Expanding agent: anhydrous calcium sulphoaluminate (10 per cent) (obtained by heating a finely-ground mixture of 50 per cent gypsum with 25 per cent bauxite and 25 per cent limestone)

Stabiliser: granulated blastfurnace slag

Expansion is due to the hydration of anhydrous sulpho-aluminate into the ettringite $(3CaO.Al_2O_3.3CaSO_4.30—32H_2O)$.

Another type is obtained from OPC and 5–7 per cent magnesia (made by calcining dolomite at 800–900 °C). Expansion is due to the hydration reaction: $MgO \rightarrow Mg(OH)_2$.

Expanding cements are not normally resistant to sulphate solutions, to sea water, acids, bases and alkaline carbonates. There is no BS specification yet for expanding cements but apart from a few differences (particularly the dimensional change) they are expected to have similar properties to those of ordinary non-expanding cements. They are useful for structural repair work. There is great potential in their use as non-shrinking, self-stressing or freely expanding cements, although successful applications have yet to be fully investigated and tried.

(*e*) *Oil-well cements.* These are composed of cementing materials usually containing a retarding agent:

Cementing materials include:

(*i*) Slow-setting Portland cements with a ratio Al_2O_3/Fe_2O_3 below 0·64.

(*ii*) Pozzolanic cements (obtained by grinding Portland cement clinker with pozzolana *or* mixing hydrated lime with pozzolana).

(*iii*) Coarsely-ground OPC with special retarders.

Retarding agents include:

(*i*) Some hydroxy-compounds, such as starch, cellulose, sugar, acids and salts of acids.

(*ii*) Calcium or sodium lignosulphonate.

Oil-well cements are used in the drilling of oil wells for cementing the steel lining tube to the rock formation and to seal off porous water or gas-bearing formations on the drilling path. They are pumped, in the form of a slurry, and sometimes taken to a considerable depth—6100 m or more— at temperatures up to 177 °C and pressures up to 124 N/mm². The cement slurry must remain sufficiently fluid to be pumped for up to about 3 hr. and then harden fairly quickly after setting. In addition, they must be fairly resistant to corrosive actions of sulphur gases and waters containing dissolved salts.

(*f*) *Hydrophobic cement.* This cement may be stored in damp humid conditions without deterioration or getting lumpy even for prolonged periods. It is obtained by grinding OPC clinker with a film-forming substance such as oleic acid, stearic acid, lauric acid or pentachlorophenol, which forms a water-repellent film round each cement grain. During the mixing process in the production of concrete, this protective film breaks down as a result of friction and the process of hydration takes place normally.

Although primarily used for prolonged storage, it can also be used as a masonry cement in view of its plasticity, water-retentivity and pumpability. It has strengths similar to OPC and may even possess some frost resistance because of its plasticity and air-entraining properties. It is, however, expensive to use.

AGGREGATES

Aggregates occupy 70–75 per cent of the total volume of a mass of concrete and, therefore, the properties of concrete are to a large extent dependent on the properties of the aggregates contained in them.

8. Definition. An aggregate is defined as an inert mineral filler used in Portland cement concrete (BS 882). The aggregates used in concrete vary in sizes:

(*a*) *coarse* aggregates (*e.g.* gravel) are 4·76 mm ($\frac{3}{16}$ in.) or more in size,

(*b*) *fine* aggregates (*e.g.* sand) are less than 4·76 mm in size,

(c) *silt* varies from 0·02–0·06 mm size,

(d) *clay* particles are much finer than 0·02 mm.

9. Classification of aggregates. Aggregates can be classified in many ways:

(i) Petrological (BS 812).

(ii) According to shape and texture (BS 812).

(iii) According to density (Table XVI).

(a) *Heavy aggregates.* These have a specific gravity of about 4·0 or more.

(i) *Magnetite.* A natural iron ore, which can be represented

TABLE XVI: CLASSIFICATION AND PROPERTIES OF AGGREGATES

	Specific gravity	Crushing strength (N/mm²)	Porosity (%)	Density of concrete (kg/m³)	Compressive strength of concrete (N/mm²)
HEAVYWEIGHT					
Magnetite (Fe_3O_4)	4·9–5·2			2900	
Barytes ($BaSO_4$)	4·50			3520	
Iron shots	7·85			5500	
Ilmenite ($FeTiO_3$)	4·68–4·76				
Haematite (Fe_2O_3)	4·9–5·3				
Chrome ore (chromite)	4·5–5·1				
Steel balls	7·85				
NORMAL WEIGHT					
Natural:					
Sand	2·65				
Gravel	(2·65)				
Crushed rocks, *e.g.*					
granite	2·6–3·0	185	0·4–3·8		
quartzite	2·6–2·7	330	1·9–15·1		
basalt	2·6–3·0	200			
sandstone	2·14–2·36				
limestone	2·5–2·8	165	0–37·6		
Artificial:					
Broken brick	1·4–2·2				
Air-cooled blastfurnace slag	2·0–3·9				
LIGHTWEIGHT					
Natural:					
Pumice	0·50–0·88			640–1440	2·0–14·0
Sawdust					
Asbestos	2·0–2·8				
Air					
Artificial:					
Furnace clinker	0·72–1·04			1040–1520	2·0–7·0
Foamed slag	0·32–0·88			960–2000	2·0–24·0
Expanded clay } Expanded shale}	0·32–1·04			720–1760	2·0–62·0
Expanded perlite	0·08–0·3			400–1120	0·5–7·0
Sintered pulverised fuel ash	0·64–0·96				
Exfoliated vermiculite	0·065–0·20			400–800	0·7–3·5

chemically as a mixed iron oxide (Fe_3O_4). It can be used as a coarse or a fine aggregate in concrete mixes. It has a high water absorption and a tendency to dust. Concrete with densities 3000–3900 kg/m³ can be readily obtained.

(ii) *Barytes* ($BaSO_4$). A barium ore (specific gravity 4·1). Density of the concrete varies with mix proportions, a value of 3700 kg/m³ can be readily obtained.

(b) *Normal weight aggregates*. These have a specific gravity normally ranging from 2·5–3·0.

(i) *Sands and gravels*. Obtained naturally from stream deposits, glacial deposits and sand dunes. Stream deposits prove to be the most satisfactory because individual particles are rounded, while weaker materials tend to be removed by abrasion. The division between sands and gravels is based arbitrarily on size: sands pass through a 4·76 mm BS test sieve while gravels are larger-sized particles retained on a 4·76 mm sieve.

(ii) *Granites and basalts*. Natural rocks which make excellent aggregates due to their hardness and toughness.

(iii) *Sandstones*. Hard and dense types make suitable aggregates. Those containing calcium carbonate are likely to be affected by acid gases in the atmosphere.

(iv) *Limestones*. Hard and dense types are suitable for use as aggregates. The softer and more porous are unsuitable because of poor abrasion and poor frost resistance.

(v) *Broken brick*. Used as aggregate provided it is free from traces of plaster which may delay the set of cement. The soluble sulphate content in the brick must not be more than 0·5 per cent (SO_3) to avoid any risk of sulphate attack. Depending on the porosity of the aggregate the density of the resulting concrete is about 2080–2240 kg/m³, which is rather light.

(vi) *Air-cooled blastfurnace slag*. Must be stable and dense (density over 1250 kg/m³). Can be used in plain, reinforced and precast concrete, but not suitable for use with HAC. Its use as coarse aggregate is dealt with in BS 1047.

(c) *Lightweight aggregates*. These are covered by the following British Specifications:

BS 877:1967 Foamed slag aggregate.
BS 1165:1966 Clinker aggregate.
BS 3681:1973 Methods for the sampling and testing of lightweight aggregates.
BS 3797:1964 Lightweight aggregates.

Lightweight aggregates are used to yield concretes usually of density less than 1920 kg/m^3. Lightweight aggregate concretes are dealt with in **16** below (*see also* Table XVI).

10. Properties of aggregates (normal weight). Ideally, aggregates should be chemically inert, durable, hard and tough and able to compact into a dense mass as well as to provide good bonding with cement paste.

In practice, aggregates vary considerably in properties, chemically and physically.

(a) *Chemical properties.* Most natural aggregates are contaminated with impurities which may be harmful to concrete.

Contaminations	Possible effect
1. *Soluble salts*	
sodium chloride	Corrosion of reinforcement and efflorescence
calcium sulphate	Set retarder and sulphate attack
organic compounds	Setting of cement affected
2. *Reactive silica* (*e.g.* in opaline or chalcedonic cherts, siliceous limestones, phyllites)	Reaction with the alkali constituents of cement (alkali-aggregate reaction)
3. *Clay and silt particles*	Bonding of the aggregates by the cement paste may be affected
4. *Weak or unsound particles* (*e.g.* coal, mica, pyrites, shale)	Strength and/or appearance affected

(b) *Physical properties.* The most important physical characteristics which have a bearing on concrete are:

(i) *Strength.* It is true to say that strong aggregates make strong concretes. Most normal weight aggregates are sufficiently strong and have good resistance to impact, crushing and abrasion.

(ii) *Porosity.* This is associated with strength as well as with water absorption, permeability and hence durability. Pores within an aggregate vary in size and those less than 4 μm in size are generally believed to affect the durability of aggregates subjected to freeze–thaw cycles.

(iii) *Thermal properties.* Aggregates have a wide range of

thermal properties. Most aggregates have coefficient of thermal expansion between 5×10^{-6} and 13×10^{-6} K^{-1} and Portland cement paste normally between 11×10^{-6} and 16×10^{-6} K^{-1}. Differential thermal expansion, if greater than $5 \cdot 5 \times 10^{-6}$ K^{-1}, may disrupt the bonding between the coarse aggregates and the cementing paste.

The other thermal properties (specific heat capacity and thermal conductivity) are important in mass concrete or where thermal insulation or fire resistance is required.

(*iv*) *Bond characteristics.* The bonding between the aggregate and the cementing paste is also affected by the shape and surface texture of the particles and by the surface coatings, *e.g.* clay, silt, dust. A stronger bond is obtained with a rougher surface texture than a smooth one. Elongated and flaky aggregates result in strength reduction by producing planes of weakness, whereas rounded and angular aggregates increase the strength of concrete. Surface coatings reduce the efficiency of bonding where coatings of organic origin may affect the hydration of cement.

(*v*) *Moisture movement* (or drying shrinkage). Due to their porous nature, aggregates expand on wetting and contract on drying. The volume changes resulting from alternate cycles of wetting and drying can create sufficient internal stress to cause deterioration of concrete, particularly thin concrete products and reinforced concrete.

Shrinkable aggregates (*e.g.* dolerites, basalts, limestones, sandstones, mudstone and gravels derived from them) have unusually high moisture movement: 0–$0 \cdot 07$ per cent for most dolerites, $0 \cdot 1$–$0 \cdot 8$ per cent for some sandstones, up to $0 \cdot 1$ per cent for some argillaceous limestones). For a dense concrete the drying shrinkage is of the order of $0 \cdot 03$–$0 \cdot 04$ per cent (with non-shrinking aggregates) and of the order of $0 \cdot 1$ per cent or more (with shrinkable aggregates).

(*vi*) *Density. Solid* density is determined in the powder form by the use of a pycnometer or a specific gravity bottle method. It refers to the volume of solid (powder) excluding pores or voids whereas *bulk* density refers to the volume of the packing including pores or voids. Consequently, bulk density is always less than the solid density and it is dependent on the efficiency of packing and hence on the size distribution (grading) and shape of particles. The percentage porosity or voids can be calculated from the relation:

Percentage porosity

$$= \frac{\text{Solid density} - \text{Bulk density}}{\text{Solid density}} \times 100$$

11. Grading of aggregates. Grading refers to the size distribution of the aggregates. It is obtained by sieve analysis which involves sieving the aggregates through a nest of standard BS sieves (BS 410) and calculating the percentage by weight passing the various sieves.

A practical result of a sieve analysis is tabulated together with the grading curves, in Fig. 4.

BS Sieve	COARSE AGGREGATE				FINE AGGREGATE			
	Weight retained (g)	Percentage retained	Cumulative (% retained)	Cumulative (% passing)	Weight retained (g)	Percentage retained	Cumulative (% retained)	Cumulative (% passing)
1½ in. (38·1 mm)	0	0	0	100	0	0	0	100
¾ in. (19·05 mm)	15	3	3	97	0	0	0	100
⅜ in. (9·52 mm)	285	57	60	40	0	0	0	100
³⁄₁₆ in. (4·76 mm)	175	35	95	5	5	2	2	98
No. 7 (2·40 mm)	25	5	100	0	20	8	10	90
No. 14 (1·20 mm)	0	0	100	0	37·5	15	25	75
No. 25 (600 μm)	0	0	100	0	75	30	55	45
No. 52 (300 μm)	0	0	100	0	62·5	25	80	20
No. 100 (150 μm)	0	0	100	0	27	18	98	2
Passing No. 100					23			
TOTAL	500				250			

(a) Sieve analysis.

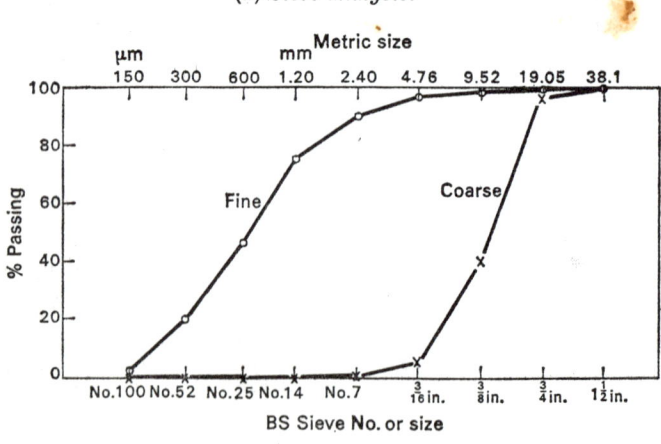

(b) Grading curves.

Fig. 4.—*Sieve analysis and grading curves for aggregates.*

12. Tests on aggregates.

(a) *Field settling test* (for clay, silt and fine dust content). Sand is poured into a 250 ml measuring cylinder containing 50 ml of an approximately 1 per cent aqueous solution of common salt until the sand level reaches the 100 ml mark. More salt solution can be added, if necessary, to bring the liquid level to 150 ml mark. Stopping the cylinder with the palm of the hand, the cylinder is shaken vigorously and allowed to stand for 3 hr. The volume of silt deposited on the surface of the sand can be measured and expressed as a volumetric percentage. According to BS 882, if the volumetric content exceeds 8 per cent, more accurate gravimetric tests must be carried out, such as by the sedimentation method described in BS 812.

The content of clay, silt and fine dust particles is limited by BS 882:

Crushed stone sand: maximum 15 per cent by weight.
Natural or crushed gravel sand: maximum 3 per cent by weight (and maximum 8 per cent by volume).
Coarse aggregate: maximum 1 per cent by weight.

(b) *Organic impurities*. The colorimetric test is now replaced by a *pH* test, which is carried out by measuring the *pH* values of cement mortars under standard conditions. If the *pH* value is less than 12·40, more consideration will have to be given to whether the type of organic impurities present in aggregates may affect adversely the hydration of the cement paste.

(c) *Moisture content*. The moisture present in a porous aggregate may be represented diagramatically, as in Fig. 5.

Free moisture (or moisture content) can be determined in the laboratory using a Pycnometer:

$$\text{moisture content} = \left(\frac{C}{A-B} \cdot \frac{(\rho-1)}{\rho} - 1 \right) \times 100$$

where A = weight of Pycnometer + sample + water to fill

B = weight of Pycnometer + water to fill

C = weight of moist sample

ρ = apparent specific gravity of aggregate on a saturated and surface-dry basis or alternatively by means of a speedy moisturemeter which makes use of a chemical (calcium carbide) to react with the free moisture to form a gas (acetylene) which is measured and the meter reading expressed directly as a percentage of wet weight. This "speedy" method is quick, compact and portable.

Total moisture content can be obtained by heating a known weight of moist aggregate at 105 °C to constant

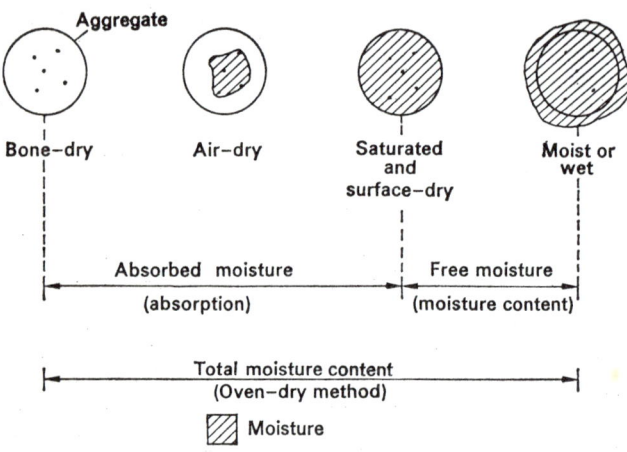

Fig. 5.—*Moisture present in a porous aggregate.*

weight, expressed as a percentage of either dry weight or wet weight.

(d) *Bulking of sand.* The volume of a given mass of sand is dependent on the moisture content. The volume is a minimum when the sand is either dry or inundated. The maximum bulking or increase in volume is between 3–10 per cent moisture content.

If V_1 = initial volume of moist sand,

V_2 = volume of inundated sand (or dry sand),

then bulking = $(V_1 - V_2)/V_2$.

Bulking becomes important when the batching of concrete

by volume is used. In this case, a bulking factor of V_1/V_2 is used.

CONCRETE MIXES

Concrete is made up of:

Cement + water + (fine & coarse) aggregates + (admixture)

(a) *Cement + water* give the cement paste which fills the space and thereby forms a bonding between the aggregate particles.

The *water* used should be fit for drinking or should be free from suspended particles, organic matter and/or salts which might affect the hydration process of cement. If the quality of water is suspect, tests should be carried out on setting times of the cement paste and the compressive strength of concrete according to BS 3148. The water content of the mix is usually expressed in terms of the water/cement ratio (W/C).

(b) *Fine and coarse aggregates* should be properly graded and free from harmful contaminations such as fine particles of clay, silt and dust, organic impurities, salts or unsound particles of coal, pyrites or shale. The proportions are usually expressed in terms of cement: fine: coarse aggregates *e.g.* 1:2:4 = 1 part cement:2 parts fine:4 part coarse aggregate by *weight* or by *volume* (which may be either bulk or solid).

If by *bulk* volume, the state of the aggregates (whether loose or compacted, damp or dry) must be clearly specified, so that proper adjustments may be further made.

(c) An *admixture* (*see* pp. 65–8) may sometimes be added with the purpose of modifying one or more properties of the concrete. The amount added is normally small and is relative to the weight of the cement. It can be in solid or liquid form; if solid and soluble it is best dissolved in the mixing water before adding to the concrete mix.

13. Fresh concrete. The main properties of freshly-mixed concrete are:

(a) Workability.
(b) Uniformity.

(c) Consistency.
(d) Segregation.
(e) Bleeding.

(a) *Workability* can be defined as the amount of useful internal work required for full compaction (definition due to Glanville). It expresses the ease of working and is closely associated with the *uniformity* or the *consistency* of the concrete mix.

(b) *Uniformity* is merely a degree of homogeneity or a state of distribution within the mix. There is often a certain degree of non-uniformity within a given batch or from one batch to another.

(c) *Consistency* describes the state of wetness of the concrete or the state of firmness of form.

(d) Owing to the heterogeneity and complex nature of the concrete mix, there is often a tendency of the heavier particles to separate from the lighter ones, *e.g.* through gravity. This form of separation, which leads to a state of non-uniformity of the mix, is known as *segregation*.

(e) One form of segregation, *e.g.* in very wet mixes where there is a tendency for the surplus water to separate and rise to the surface, is known as *bleeding or water-gain*. Bleeding often involves the formation of a scum or "laitance" on the surface.

(f) Table XVII summarises the factors affecting workability of fresh concrete. No absolute test is as yet available for *measuring workability*. However, the assessment tests mostly used in the U.K. are the Slump test, the Compacting factor test (both of which are described in BS 1881) and the Vebe consistometer test.

(i) *Slump test.* This is a simple method most commonly used on the site for quality control test of a concrete mix. It is more appropriate for testing consistency and uniformity than workability. It is suitable for wet mixes but not for very wet or very dry mixes. The result of this test can be very subjective, as it involves filling a standard slump cone with the concrete mix compacted by hand tamping (according to BS 1881) and measuring the amount of slump (in mm) after the cone has been carefully removed. The greater the sagging or slump the greater the workability. It also gives indication of the cohesiveness and uniformity of the mix from the form of slump obtained, *i.e.* normal slump, shear slump or collapse.

TABLE XVII: FACTORS AFFECTING WORKABILITY

Factors	Effect[1]
1. Cement	
(i) Type	workability $\nearrow$ [2] for OPC, $\searrow$ for RHPC (finer)
(ii) Amount	workability $\nearrow$ as amount $\nearrow$ (e.g. very rich mixes)
2. Aggregates	
(i) Size	workability $\nearrow$ as size $\nearrow$
(ii) Grading	workability $\nearrow$ as grading $\nearrow$ (i.e. coarser)
(iii) Shape	workability $\nearrow$ for rounded but $\searrow$ for angular aggregate
(iv) Texture	workability $\nearrow$ for smooth, but $\searrow$ for rough aggregate (concept of surface area)
3. Water content	workability $\nearrow$ as water content $\nearrow$
4. Consistency	workability $\nearrow$ as consistency $\nearrow$ (since consistency $\nearrow$ as water content $\nearrow$)
5. Air-entrainment	workability $\nearrow$ as air content $\nearrow$ (air acting as lubricant)
6. Water-reduction (water-reducing agents)	workability $\nearrow$ as water reduction $\nearrow$
7. Finely-divided materials (e.g. pozzolans)	workability normally improved

[1] When considering the effect of one particular factor, all other factors are assumed to be fixed.

[2] $\nearrow$ increases; $\searrow$ decreases.

(ii) *Compacting factor test.* Mixes of medium consistency, which may be too dry to give a measurable slump, are best tested by this method, which is based on the extent of compaction of a mix from its free fall over a standard height. The mix is carefully transferred to hopper A without any compaction, then allowed to fall into hopper B and finally into cylinder C (Fig. 6). The partially-compacted weight of the concrete (W_1) filling the cylinder (after removing the concrete above the rim of the cylinder) is found. The cylinder is emptied and filled with concrete in layers of approximately 50 mm, each layer being compacted by hand or by vibration. The weight of the fully compacted concrete (W_2) is then found. The compacting factor (C.F.) is obtained by the ratio W_1/W_2.

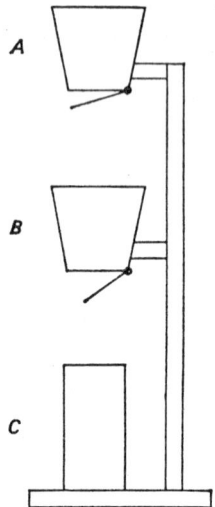

FIG. 6.—*Compacting factor apparatus.*

C.F. approaches the value of unity as the workability of the mix increases, but never exceeds unity.

This test is more reliable than the slump test and is most suitable for site mix.

(*iii*) *Vebe consistometer test.* This is suitable for dry mixes which are to be compacted by vibration. The mix is filled into a slump cone (as in the slump test) which is inside a hollow

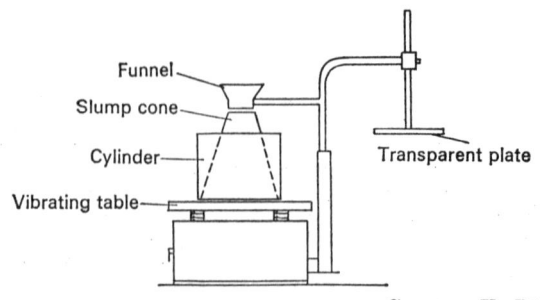

Courtesy V. Bährner

FIG. 7.—*Vebe consistometer.*

cylinder (Fig. 7). After the slump cone is removed, a loaded transparent circular plate is rested on the concrete cone. By vibrating the cylinder, the mix is compacted from a conical to a cylindrical shape by the combined action of the vibration and the weight of the loaded plate and the time taken for this change in shape is noted as Vebe seconds. The higher the value of Vebe seconds the lower is the workability of the mix.

Table XVIII shows some comparative values for workability of concrete for different purposes.

TABLE XVIII: WORKABILITY FOR DIFFERENT PURPOSES

Purpose	*C.F.*	*slump* (mm)[1]	*Vebe* (seconds)
Very high-strength concrete for prestressed concrete sections compacted by heavy vibration	0·70–0·78	0	over 20
High-strength concrete sections, pavings and mass concrete compacted by vibration	0·78–0·85	0–25	7–20
Normally reinforced concrete section compacted by vibration Hand-compacted mass concrete	0·85–0·92	25–50	3–7
Heavily-reinforced concrete sections compacted by vibration Hand-compacted concrete in normally-reinforced slabs, beams, columns and walls	0·92–0·95	50–100	1–3
Heavily-reinforced concrete sections compacted without vibration, and work where compaction is particularly difficult	over 0·95	100–150	0–1

[*Concrete practice—Cement and Concrete Association.*
[1] Approx. metric conversions.

14. Hardened concrete. The main properties of hardened concrete are strength and durability. Other properties include elasticity, permeability (or water tightness), drying shrinkage, creep and thermal properties.

Strength is the resistance to applied forces and, in the case of concrete, is considered as a criterion of quality. In general, stronger concretes are denser, more watertight and more durable but more liable to cracking owing to higher drying shrinkage. Strength is measured in units of stress (force per unit area) *e.g.* in N/mm².

(a) *Types of strength*

(i) *Compressive and tensile strengths.* Plain concrete structures are strong in compression and weak in tension, hence the need for steel reinforcement in concrete. There is no simple relation between them but as an approximation the tensile strength can be regarded as 7–11 per cent of compressive strength.

(ii) *Flexural strength.* This is expressed in terms of "modulus of rupture," which is the maximum transverse breaking stress, applied under specified conditions, that the concrete specimen can withstand before rupture.

The modulus of rupture (M) of a test piece of rectangular cross-section is given by:

$$M = \frac{3\ W\ l}{2\ b\ d^2},$$

where W = breaking load,

l = distance between knife-edges in the transverse-strength test,

b = breadth of the test piece, and

d = depth.

For a cylindrical test piece of diameter D, the equation is:

$$M = \frac{8\ W\ l}{D^3}.$$

Flexural strength is usually higher than the tensile strength and, as an approximation, is 11–23 per cent of compressive strength.

(iii) *Shear strength.* Shear stress is always accompanied by tensile and compressive stresses. It is about 12 per cent of compressive strength, sometimes even 50–99 per cent of compressive strength.

(iv) *Bond strength* (with reinforcement). This is a measure of the resistance to slipping of steel reinforcing bars in concrete and is dependent on various factors such as type of cement, W/C ratio, type of reinforcement, etc.

(b) *Factors affecting strength.*

(i) *Constituents of concrete.*

I. *Cement:* due to chemical composition and fineness, *e.g.* C_3S is stronger than C_2S at early ages, C_3A shows greater strength at very early age, greater strength using finer type of cement.

II. *Aggregates:* due to type, shape and texture, *e.g.* normal dense aggregates are generally much stronger than the cement paste, whereas lightweight aggregates are relatively weak. Stronger concrete is obtained using angular or rough rather than rounded or smooth aggregates (assuming constant W/C ratio).

III. *Water:* depending on quality, *e.g.* harmful contaminations may affect the setting and hardening process of the cement paste. Sea water, if used, may reduce the compressive strength of concrete at 28 days by about 12 per cent.

IV. *Admixtures:* depending on type, *e.g.* air-entraining agents lower strength as a result of air-entrainment, accelerators (*e.g. Ca* formate) increase strength at an early age, water-reducing agents may increase strength by reducing the water content.

(*ii*) *Mix proportions and quality control.* A denser and stronger concrete is obtained by using well-designed mixes, well-graded aggregates, close control and supervision. The choice of W/C ratio required for workability of the mix is an important factor.

(*iii*) *Curing conditions.* The purpose of curing is to allow the hydration process to complete its course, which at normal temperature may last up to 30–50 years. Consequently greater strength is achieved by allowing a longer period of moist curing. The strength of concrete can be improved by raising the curing temperature. However, strength is reduced when destructive agencies are present during the curing process.

(*c*) *Testing of hardened concrete.*

(*i*) *Compressive strength test for concrete cubes.* Concrete cubes are cast in 101 mm (4 in.) or 152 mm (6 in.) steel cube moulds according to whether the maximum size of aggregates is up to 19 mm (0·75 in.) or 38 mm (1·5 in.) or less respectively. They are demoulded after curing for 24 hr. in humid conditions at normal room temperature, and stored in water until tested in a compression machine either wet or after drying in oven at 105 °C. The method of this test is described in BS 1881: Part 4: 1970—Method of testing concrete for strength.

(*ii*) *Non-destructive methods of testing.* The compression test is a destructive method. The test specimen, once tested destructively, cannot be used again. This limits the investigation of progress of strength development in relation to aging, and the assumption that test specimens prepared under the same controlled conditions with exactly the same mix proportions are identical in properties is simply not true at all.

The advantage, therefore, of the non-destructive methods of testing is obvious in this respect. Two such methods in common use are the *Rebound test* and the *ultrasonic test*.

I. *Rebound test (Schmidt Rebound Hammer)*. This consists of a small spring-loaded hammer which essentially measures the surface hardness rather than the strength of the concrete. It is simple to use, but the results can be questionable. The values given in "rebound numbers" can be calibrated and converted into compressive strength values. It can be useful for quality control, *e.g.* for checking the uniformity of a concrete structure. The result can be affected by various factors such as the size of the test specimen, the type of surface (whether smooth or not so smooth, flat or curved, dry or wet) and the position of use (*i.e.* whether tested vertically or horizontally).

II. *Ultrasonic test*. This measures the electrodynamic modulus of elasticity (*E*) of the test specimen (in the form of a regular prism).

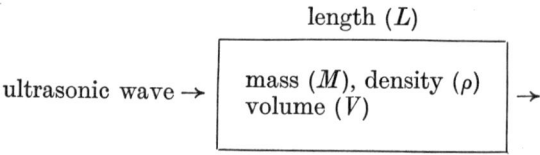

By passing an ultrasonic wave through a test specimen of length (*L*), mass (*M*), volume (*V*) and density (*ρ*), the velocity of the wave (*v*) is given by the relation:

$$v = \sqrt{\frac{E}{\rho}}$$

where *E* = modulus of elasticity of material of test
        specimen,

and ρ = density of specimen = $\dfrac{M}{V}$.

By measuring the velocity (*v*) from $v = \dfrac{L}{t}$

where   *t* = time taken for the wave to travel along
        the distance *L*,

E can be evaluated, *i.e.*

$$E = v^2\rho = \frac{L^2 M}{t^2 V}.$$

15. Manufacture of concrete. The various stages include:

(*a*) Batching.
(*b*) Mixing.
(*c*) Transporting.
(*d*) Placing.
(*e*) Compacting.
(*f*) Curing.
(*g*) Finishing.

(*a*) *Batching.* The process of measuring the constituents of the concrete is known as *batching*. Batching by weight is preferred to batching by volume. Cement is more conveniently batched by bags of 50 kg. For aggregates, the moisture present (whether as free moisture or absorption) must be known and proper adjustment made to the water requirement, remembering that the *W/C* ratio is computed on the basis of satured surface-dry aggregates. In addition, the bulking of sand must be allowed for in the case of volume batching. Strict control is required for the measurement of water, and particularly of admixtures.

(*b*) *Mixing.* Its purpose is to produce uniformity and homogeneity of the concrete mix. Mixing can be done manually (for small jobs) or mechanically (by using power-operated mixers). Mixing times vary according to the size of the mix and the type of mixer used (approximately 1–10 min.).

Two types of drum mixers—Tilting and Non-tilting types—are covered by BS 1305.

It is worth noting that under-mixing is just as bad as over-mixing.

(*c*) *Transporting or conveying.* After mixing, the concrete is to be transported from the mixer to the formwork. The various means of transporting include the use of:

(*i*) batch containers (wheelbarrows, lorries, skips);
(*ii*) displacement pumps (involving pumping concrete through steel pipelines);
(*iii*) pneumatic method (which involves the use of compressed air for pumping concrete through pipelines);

  (iv) chutes and conveyor belts (proper use of downpipes is essential).

Consistency of the mix should be properly controlled, as pumpable concrete usually requires higher consistency. Strict precautions must be taken to avoid segregation taking place during the transporting stage.

(d) *Placing.* The placing of the concrete in position (*e.g.* in formworks) follows soon after transporting. Care should be taken to avoid segregation occurring and the formation of unnecessary construction joints. Dropping the concrete from a considerable height increases the risk of segregation.

(e) *Compacting.* In order to obtain a dense homogeneous mass all entrapped air and voids must be removed by compaction or consolidation. This can be achieved by either hand tamping or vibration. The general types of vibrators in use include:

  (i) Internal or Immersion vibrators (should not be brought too close to formwork or reinforcement).

  (ii) External vibrators (including pneumatic and electrical hammers applied to but not clamped to formwork).

  (iii) Surface vibrators (including beam vibrators or pan vibrators).

  (iv) Table vibrators (mainly used for compacting small precast concrete units).

(f) *Curing.* After placing and compacting, the concrete must be allowed to set and harden for an appropriate period under controlled humidity and temperature. This is the curing process. The curing period varies, usually from 1–7 days depending on the type of cement used and the ambient conditions of temperature and humidity.

Steam curing is usually used in mass production of precast concrete units in order to speed up the hydration and hardening process. Steam curing can be carried out at atmospheric or high pressures. Curing methods include:

  (i) Use of waterproof membranes to prevent evaporation of water from the mix (*e.g.* waterproof papers, polythene sheets).

  (ii) Use of damp or wet coverings (*e.g.* hessian) to provide continuously a uniform film of moisture over the concrete.

  (iii) Direct use of water by ponding or sprinkling. Ponding is suitable only for flat concrete surfaces such as floor and roofs.

(*iv*) Use of sealing compounds (*e.g.* bituminous emulsions and cutbacks) applied all over the surface of the concrete to prevent water loss.

(*g*) *Finishing.* After the formwork is removed, the concrete surface may require additional attention such as filling up the bolt holes, smoothing any surface irregularities, patching up any defective surface or treating the surface to obtain a desired finish such as the exposed aggregate finish. The latter can be achieved *in situ* by any of the following methods:

(*i*) Early removal of formwork followed by brushing and washing.

(*ii*) Use of retarders painted on formwork followed by brushing and washing.

(*iii*) Tooling, using a mechanical bush hammer, a point tool or a comb chisel.

(*iv*) Sand-blasting the concrete surface.

SPECIAL TYPES OF CONCRETE

16. Lightweight concretes. These have low densities ranging from 320–1920 kg/m³ (normal dense concretes have densities 2240–2600 kg/m³). The higher density range can be used for load-bearing purposes and the lower density range for thermal insulation purposes. They can be easily worked using ordinary woodworking tools. Their use is gaining popularity, as savings in weight, time and expense for erection and handling are possible.

Lightweight concretes can be produced by three different methods:

(*a*) By using lightweight aggregates (to produce lightweight aggregate concretes).

(*b*) By introducing gas or air bubbles (to produce aerated or cellular concretes).

(*c*) By leaving out the fine aggregate from the mix (to produce "no-fines" concretes).

(*a*) *Lightweight aggregate concretes.* Constituent materials are:

Binder: Portland cement;
Fine aggregate: sand or lightweight aggregate (grading of less than 5 mm);

Coarse aggregate: lightweight aggregate (grading limits 19–5 mm);
Water: drinkable.

The properties of lightweight aggregate concrete are mainly dependent on the type and grading of aggregates, the mix proportions and the water content.

The characteristic properties are:

(*i*) low density and low strength,
(*ii*) high porosity,
(*iii*) good thermal insulation and fire resistance,
(*iv*) high moisture movement and high initial drying shrinkage, and
(*v*) good sound absorption.

Main applications include:

(*i*) *Structural uses:* multi-storey buildings, precast wall panels, precast floor and roof units (BSS CP 116: 1965—The structural use of precast concrete; CP 111: 1964—Structural recommendations for load-bearing walls.

(*ii*) *Blockwork* (BSS 2028, 1364: 1968—Precast concrete blocks): for production of load-bearing and non-load-bearing concrete building blocks.

(*iii*) *Insulation:* thermal insulation of roof and floor screeds.

(*b*) *Aerated concretes* (gas or foamed concretes). More appropriately called aerated mortars, since the coarse aggregates are not normally used.

Constituent materials include:

Binder: cement or lime—PBFC and HAC are not suitable for autoclaved aerated concretes;

Fine aggregate: sand or lightweight aggregate (such as pfa);

Water: drinkable;

Aerating agent: chemical agent (*Al*, *Fe*, or *Zn* powder, $H_2O_2 + CaOCl_2$, CaC_2); foaming agent (glue resin, saponified resin, or alumino-sulphonate).

These concretes are characterised by:

(*i*) very low densities, strengths and thermal conductivities, good nailability,
(*ii*) high fire resistance.

Main applications include:

(i) *Cast in situ:* (foamed concrete) for insulating floor and roof screeds.

(ii) *Precast units:* blocks, roof and floor slabs, wall slabs, partition slabs and lintels.

(c) *"No-fines" concretes.* Constituent materials include:

cement and coarse aggregate (which may be gravel, crushed stone or lightweight aggregate).

Characteristic properties:

(i) low-drying shrinkage (about 0·02 per cent, which is half the value for ordinary dense concrete);

(ii) density: 1600–2000 kg/m³ with normal aggregates, as low as 640 kg/m³ with lightweight aggregates;

(iii) thermal insulation improved with lightweight aggregates;

(iv) high water absorption;

(v) not easy to nail or cut.

Main applications include:

(i) Infill panels and partitions.

(ii) Blockwork.

(iii) Roadwork and road surfacings (to improve water drainage and skid resistance).

17. Heavyweight concretes (or high-density concretes). These have densities greater than 2600 kg/m³ and are produced using high-density aggregates (BS 4619:1970—Heavy aggregates for concrete and gypsum plaster).

High-density aggregate	Density of aggregate	Density of concrete[1]
Magnetite (Fe_3O_4)	4900–5100 kg/m³	2900 kg/m³
Barytes ($BaSO_4$)	4100–4500 kg/m³	3520 kg/m³
Iron shot or scrap iron	7850 kg/m³	5500 kg/m³

[1] Values vary with mix proportions.

Mix proportions: Generally, cement: fine and coarse aggregate = 1:5 to 1:9 (by weight) with W/C ratio 0·5–0·65.

Main properties:

(*i*) high density;
(*ii*) owing to high density of aggregates used, segregation tends to occur during placing.

Main applications:

(*i*) X-ray and nuclear radiation shielding.
(*ii*) Biological shields for nuclear reactors.
(*iii*) Ballast blocks for ships.
(*iv*) Balance weights for lift bridges.

18. Reinforced concrete. The problem of the low tensile strength of concrete can be solved by use of reinforcement (such as steel rods or bars, glass or carbon fibres) embedded in the concrete. The result is a composite reinforced material which can resist high compressive and tensile stresses. With steel reinforcement, protection from corrosion is an important factor and can be achieved with a sufficiently thick surrounding layer of dense concrete and/or a surface coating applied to the steel rods or bars.

19. Prestressed concrete. This is reinforced concrete using prestressed steel bars or wires. It is usually made by applying a tensile stress to high-carbon steel bars or wires before the concrete is allowed to harden. As a result, a much stronger concrete is obtained.

Specifications for and use of reinforced and prestressed concretes are dealt with in the following British Standards:

BS 2539 : 1954 Preferred dimensions in reinforced concrete structural members.
BS 2691 : 1969 Steel wire for prestressed concrete.
BS 3617 : 1971 Seven-wire steel strand for prestressed concrete.
BS 4449 : 1969 Hot rolled steel bars for the reinforcement of concrete.
BS 4461 : 1969 Cold worked steel bars for the reinforcement of concrete.
BS 4466: 1969 Bending dimensions and scheduling of bars for the reinforcement of concrete.
BS 4482 : 1969 Hard-drawn mild steel wire.
BS 4483 : 1969 Steel fabric for the reinforcement of concrete.

BS 4486 : 1969 Cold worked high-tensile alloy steel bars for prestressed concrete.

BS 4975 : 1973 Prestressed concrete pressure vessels for nuclear reactors.

CP 110   : The structural use of concrete.
  Part 1: 1972 Design, material and workmanship.
  Part 2: 1972 Design charts for singly reinforced beams, doubly reinforced beams and rectangular beams.
  Part 3: 1972 Design charts for circular columns and prestressed beams.

CP 114   : Structural use of concrete in buildings.
  Part 2: 1969 Metric units.

CP 115   : The structural use of prestressed concrete in buildings.
  Part 2: 1969 Metric units.

CP 117   : Composite construction in structural steel and concrete.
  Part 1: 1965 Simply-supported beams in building.
  Part 2: 1967 Beams of bridges.

CP 2007 : Design and construction of reinforced and prestressed concrete structures for the storage of water and other aqueous liquids.
  Part 2: 1970 Metric units.

ADMIXTURES

An admixture is often a chemical substance (other than coarse or fine aggregates, cement or water) which is added in small amounts just before or during the mixing stage of concrete production. The purpose of using the admixture is mainly to give a desired and beneficial modification of the behaviour of concrete in the freshly-mixed and/or hardened state. Chemical admixtures are generally highly complex substances and the numerous proprietary brands on the market often contain more than one type of admixture, so that great care and consideration should be taken in their use. If in doubt, the manufacturers should be consulted. The misuse or abuse of admixtures is detrimental. The dosage used can be very critical.

TABLE XIX: ADMIXTURES—FUNCTIONS, MECHANISMS AND EFFECTS

Type	Functions	Mechanisms	Effects on fresh concrete	Effects on hardened concrete
Accelerators	To speed up: (a) Setting (b) Rate of early strength development	By neutralising the retarding effect of gypsum on C_3A	(i) Setting time ↘ (ii) Workability ↗ (especially AE) (iii) Air content ↗ (AE) (iv) Bleeding ↗	(i) Frost resistance ↗ (ii) Corrosion risk ↗ (especially prestressed steel) (iii) Drying shrinkage ↗ (iv) Strength ↗ (especially at early ages)
Retarders	To slow down: (a) Setting (b) Rate of strength development	Adsorption mechanism: by forming a protective coating around C_3A particles	(i) Setting time ↗ (ii) Workability ↗ (especially AE) (iii) Air content ↗ (AE)	(i) Frost resistance ↗ (ii) Shrinkage ↗ (for some) (iii) Strength ↗ (especially WR)
Water-reducers (WR)	To reduce water requirements of concrete for a given consistency	The admixtures are adsorbed on cement particles, resulting in a better dispersion (repulsion)	(i) Setting time (depending on type) (ii) Workability ↗ (iii) Air content (depending on type) (iv) Bleeding (AE) ↗ or (NAE) ↗ (depending on type) (v) Water reduction (vi) Slump loss	(i) Strength (depending) ↗ (ii) Shrinkage (depending on type) ↗ (iii) Frost resistance ↗ (iv) Modulus E and bond strength improved ↗ (v) Creep (vi) Resistance to sulphates improved ↗

Admixture	Purpose	Mechanism	Effects on fresh concrete	Effects on hardened concrete
Air-entraining agents (AE)	To entrain air: (i) Freeze/thaw resistance ↗ (ii) Workability ↗	AE agents are surface active agents. Minute bubbles are formed which are uniformly distributed throughout concrete	(i) Setting time (depending) (ii) Workability ↗ (iii) Air content ↗↘ (iv) Bleeding and segregation ↗ (v) Water-reduction (depending)	(i) Freeze/thaw resistance ↗ (ii) Strength (depending)
Workability aids (WA)	To improve workability	Like AE, are surface-active agents. Involved in lowering initial resistance, frictional resistance and viscous resistance	(i) Workability ↗ (ii) Water-reduction ↗↘ (iii) Air content (depending) (iv) Bleeding and segregation ↗ (v) Setting (depending)	(i) Strength ↗ (ii) Shrinkage ↘ } because (iii) Permeability ↘ } low W/C
Mineral admixtures (MA)	(a) To improve workability (b) To reduce heat rise (c) To cheapen cost	The very fine particles are incorporated into the floccular structure of cement paste, so that more water can be held (volume and workability) ↗↗	(i) Workability ↗ (ii) Heat evolution ↗↗ (iii) Bleeding (depending)	(i) Chemical resistance ↗ (ii) Strength ↗ or ↘ (depending) (especially later age) (iii) Alkali-aggregate reaction ↗↗ (iv) Permeability ↗↗ (v) Resistance to cracking ↗

↗ increases; ↘ decreases.

[C. V. Y. Chong, Concrete admixtures, Journal of Society of Engineers, Vol. LXIII, No. 4, Oct.–Dec., 1972.]

For convenience, admixtures can be classified into various important groups:

Type of admixture	Examples
Accelerator	Na_2CO_3, Na_2SO_4, triethanolamine
Retarders	$CaSO_4$, carbohydrates (e.g. sugars), lignosulphonates
Water-reducers (WR)	lignosulphonates, hydroxy organic acids and salts, calcium formate
Air-entraining agents (AE)	wood resins and their soaps, some fats and oils and their fatty acids, some lignosulphonates
Workability acids (WA)	some WR and AE (e.g. lignosulphonates)
Mineral admixtures (MA)	finely-divided materials, e.g. pozzolanic materials, blastfurnace slag, clay
Waterproofers	sodium silicate (pore-filling), Ca soaps (water-repellent or non-wettable)
Corrosion inhibitors	sodium benzoate, sodium nitrite
Fungicidal, germicidal and insecticidal admixtures	halogenated phenols, copper compounds

Table XIX summarises the functions, mechanisms and effects on concrete for some of the important admixtures.

The BS specification on admixtures (accelerating, retarding and water-reducing) was published in 1974 (BS 5075: Part 1: 1974).

Other existing BS specifications and codes of practice are:

Use of calcium chloride (CPP 114, 115, 116, 117, 123.101, 2007, BS 3587).
Use of pfa (BS 3892).
Use of pigments (BSS 3798, 1014).
Ready-mixed concrete (BS 1926).

Performance tests have to be carried out under controlled laboratory as well as site conditions.

MIX DESIGN

The basic principle of mix design is to obtain as economically as is feasible, a concrete mix of a required quality—workability, strength and durability—for a specific job by proper selection and proportioning of cement, fine and coarse aggregates and water (and sometimes admixtures if necessary).

BS CP 114 provides for nominal mixes, standard mixes and designed mixes:

20. Nominal mixes. Batched by volume, typical mixes for reinforced concrete being $1:1:2$, $1:1\cdot5:3$, $1:2:4$ (cement:fine aggregate:coarse aggregate). Equivalent "all-in" mixes may be used for foundation mass concrete.

21. Standard mixes (also BS CP 116). Batched by weight, the fine and coarse aggregates expressed in terms of dry weight per bag (50 kg) of cement. Like nominal mixes, they are simple to use and are based on published data and charts.

22. Designed mixes. There are many methods available and the British method most commonly used is the Road Note No. 4 Method (*Road Research Laboratory—Design of concrete mixes*, H.M.S.O., 2nd edition, reprinted 1962). This method has now been replaced by *Design of Normal Concrete Mixes* by Teychenné, Franklin and Erntroy (H.M.S.O., 1975). (*See* Appendix V.)

The Road Note No. 4 method is based on grading curves for combined coarse and fine aggregates and gives useful relationships between mix proportions, W/C ratio and workability:

$$\text{Strength} \propto \frac{1}{W/C}$$
$$\text{Workability} \propto W/C$$
$$\propto C/A$$
$$\propto \text{aggregate size}$$
$$\text{Durability} \propto \text{strength} \propto \frac{1}{W/C}$$

($W/C \not> $ a specified maximum, depending on the condition of exposure).

The logical procedures of designing a dense concrete mix according to Road Note No. 4 are outlined below:

(a) *Average strength*. Mixes are usually designed on the basis of a *minimum* crushing strength at 28 days. Owing to the variation in strength caused by variations in control of quality and degree of site supervision, an *average* strength greater than the minimum strength is aimed at:

    Very good control conditions:
    Minimum strength = 75 per cent average strength
    Fair control conditions:
    Minimum strength = 60 per cent average strength
    Poor control conditions:
    Minimum strength = 40 per cent average strength

(b) *Water/cement ratio (W/C)*. The W/C ratio which gives the required *average* strength is found from the graphical relationship between strength at various ages and the W/C ratio for a given type of cement to be used.

The value of such a W/C ratio must be checked to ensure that it does not exceed a specified *maximum* for the condition of exposure to be experienced.

(c) *Workability*. The degree of workability (very low, low, medium, high) for the specified job is decided on the basis of experience and tests as well as on the conditions and methods of placing and compacting.

(d) *Aggregate/cement ratio (A/C)*. The total A/C ratio for the required workability is determined from the published data, having regard to the appropriate gradings and shapes of the aggregate available for use. The leanest and most economical mix (*i.e.* the maximum corresponding A/C value) must be chosen, but not too lean as to cause segregation under normal conditions of mixing, transporting and placing. The A/C ratio and the required grading of the aggregate (grading Nos. 1, 2, 3, 4) are then fixed.

(e) *Fine aggregate/coarse aggregate ratio*. The experimental gradings of the fine and the coarse aggregates used are found by sieve analysis. The proportion of fine to coarse aggregate to fit as closely as possible the required combined grading (found in (d) above) is determined either by a graphical or a calculation method.

(*f*) Finally, the *batch proportions* of cement, fine aggregate, coarse aggregate and water are computed. This merely gives a guide to the mix proportions, which may require further modification or adjustment after making and testing trial mixes. Workability is usually measured by the compacting factor test. It can be increased by increasing the C/A ratio or reduced by reducing the C/A ratio, the W/C ratio being kept constant always.

There are obvious limitations to this method, which is heavily based on published data which are outdated and far from comprehensive.

PROGRESS TEST 4

1. Differentiate between "mortar" and "concrete". (p. 24)

2. Outline the various stages involved in the manufacture of Portland cement. (1)

3. Describe, with the aid of chemical equations, the chemical changes which take place during the production of Portland cement. (2)

4. What are the main mineralogical constituents of Portland cement? Explain their role in influencing the properties of cement. (2, Table VIII)

5. The following gives the analytical data of a typical Portland cement: $CaO = 63{\cdot}0\%$, $SiO_3 = 20{\cdot}0\%$, $Al_2O_3 = 6{\cdot}0\%$, $Fe_2O_3 = 3{\cdot}0\%$, $MgO = 1{\cdot}5\%$, $SO_3 = 2{\cdot}0\%$, others $= 4{\cdot}5\%$.

Using Bogue's equations, calculate its mineralogical constituents. (2)*

6. Write down the hydration reactions of the main mineralogical constituents of cement and indicate the order of reaction rates in each case. (3, Table IX)

7. Account for the stages leading to the setting and hardening of Portland cement. (4)

8. List the tests on Portland cement as specified by the British Standards Institution and state the specifications recommended for both the ordinary and the rapid-hardening Portland cements. (5, Table XI)

9. Give a brief account of the composition, properties and uses of the following cements:

(*a*) OPC, (*b*) RHPC, (*c*) LHPC, (*d*) PBFC, (*e*) SRPC. (6)

10. What is meant by "conversion" in the case of high-alumina cement concrete? Explain the possible effects of conversion. (7)

11. Why is HAC resistant to chemical attack? (7(*a*))

12. How is SSC different from OPC? (7(*b*))

13. For what purpose is masonry cement used ? (7(c)

14. What is the cause of expansion in expansive cement ? (7(d))

15. What are the merits and limitations of using hydrophobic cement ? (7(f))

16. Define the term "aggregate" used in concrete technology. How may aggregates be classified ?

Give examples of the various classes of aggregates. (8, 9)

17. List some of the undesirable impurities normally found in contaminated aggregates.

What are the main objections to using contaminated aggregates in concrete ? (10)

18. What is meant by "grading" of aggregates ?

What is the main difference between continuous grading and gap grading of aggregates ? (11)

19. Outline the procedure and specifications, if any, for the following tests on aggregates:

   (a) Field settling test.
   (b) Organic impurities.
   (c) Moisture content.
   (d) Bulking of sand.   (12)

20. What are the important properties of concrete in the freshly-mixed and in the hardened state ? (13, 14)

21. Explain the term "workability" in the case of a freshly-mixed concrete. State the various factors upon which workability of fresh concrete depends. (13)

22. List the merits and limitations of the three methods of test for workability of fresh concrete. (13)

23. State the various factors affecting the strength of hardened concrete. (14)

24. Describe *one* non-destructive test for hardened concrete. (14(c))

25. What are the various stages involved in the production of concrete ? (15)

26. Mention the three different methods of producing light-weight concretes. (16)

27. List the characteristics and main uses of the three different types of lightweight concretes. (16)

28. What are the main uses of high-density concretes ? (17)

29. Why is it necessary to use reinforced and prestressed concretes for structural concrete members ? (18, 19)

30. What is an "admixture" used in concrete ?

List the various types of admixtures available and give two examples of each type. (Table XIX)

31. Differentiate between nominal, standard and designed mixes.

State the factors affecting the proportioning of a concrete mix. (20–22)

32. Outline the various steps in designing a concrete mix. **(22)**

* *Answer to Question 5:* $C_3S = 54 \cdot 1\%$, $C_2S = 16 \cdot 6\%$, $C_3A = 10 \cdot 8\%$, $C_4AF = 9 \cdot 1\%$.

EXAMINATION QUESTIONS

1. State the conditions which would make:

   (a) Ordinary Portland cement,
   (b) High-alumina cement,
   (c) Low-heat Portland cement,

a suitable choice for making concrete.

Outline the curing treatment needed in each case and give some estimate of their relative rates of strength development.

(P.S.B. B.A. Arch.)

2. Why are some contaminations objectionable in aggregates? What possible effects can they have on the properties of concrete using such contaminated aggregates?

Describe a method of test for:

   (a) Workability of a freshly-mixed concrete.
   (b) The suitability of aggregates for use in concrete.

(P.S.B. Grad. Dip. Arch.)

3. (a) Explain the term "grading curve" as applied to fine aggregates for concrete, indicating the significance of zones and zone limits.

(b) Aggregates A and B are supplied and are used to give a specified grading, C. Determine the proportions in which they must be combined.

Percentage passing	A	B	C
BS Sieve:			
No. 100	0	0	0
No. 52	12	0	5
No. 25	60	0	21
No. 14	75	0	28
No. 7	85	0	35
5 mm	90	5	42
10 mm	100	40	65
20 mm	100	90	100
40 mm	100	100	100

(P.S.B. H.N.D. Bldg.)

4. Define "permeability" and "porosity" of a building material; indicate their inter-relationship, if any.

Discuss with the aid of examples the role of permeability and porosity in the deterioration of Portland cement concrete.

(P.S.B. B.Sc. Bldg.)

5. With reference to concrete technology, describe the following operations:

(a) Batching, (b) Placing, (c) Consolidating, (d) Curing, (e) Finishing.

Why is batching by weight preferred to batching by volume?

(P.S.B. Pre-Inter. Arch.)

6. (a) What do you understand by "air-entrainment" in concrete?

(b) What are its effects on the following properties of concrete:

(i) workability, (ii) density, (iii) strength, (iv) bleeding, (v) segregation, (vi) durability?

(c) Outline a method for the determination of air content in freshly-mixed concrete.

7. (a) Explain why *batching by weight* is better than *batching by volume*.

(b) In designing an ordinary concrete mix, discuss the role played by the following variables:

(i) water-cement ratio, (ii) cement-aggregate ratio, (iii) grading of aggregate, (iv) workability of the mix.

(P.S.B. B.Sc. S.E.)

8. What is the relation between the *strength, workability* and *W/C ratio* of a concrete mix?

Describe the operations and implications of the following tests:

(a) Compacting factor test, (b) Slump test, (c) Vebe consistometer test.

(P.S.B. Pre-Inter. Arch.)

9. (a) List the factors to be considered in the production of good-quality concrete.

(b) What is the possible effect of calcium chloride additive on a concrete mix and its steel reinforcements?

(P.S.B. H.N.D. S.E.)

10. (a) Explain the principal advantage that concrete pumps have over other plant used for transporting and placing concrete.

(b) Discuss the organisation of a large concrete pour, using mobile pumps for placing.

(c) State THREE properties required of a concrete mix which would make it suitable for pumping.

(I.O.B. Assoc. Part 1)

11. (a) What are the principal factors in concrete mix design ?
  (b) Describe a procedure to be followed in designing a concrete mix.
  (c) Describe circumstances in which design of concrete mixes is most effective.

(I.O.B. Final Part 1)

12. What are the advantages and disadvantages of using admixtures in concrete ?
In particular, discuss the uses and limitations of calcium chloride as an admixture.

(B.S.B. B.Sc. S.E.)

13. "The design of lightweight concrete mixes differs considerably from that of the normal dense concrete mixes." Discuss the differences implied and the necessary steps to be taken with reference to:

(a) No-fines concrete.
(b) Lightweight aggregate concrete.
(c) Aerated concrete.

(B.S.B. B.Sc. S.E.)

14. What are the main advantages of "non-destructive" testing of hardened concrete ? Outline the procedures and implications of the following methods:

(a) Electrodynamic test for elasticity of concrete.
(b) Schmidt-hammer test for strength of concrete.

(B.S.B. B.Sc. S.E.)

15. Describe the important factors which govern the proportioning of a concrete mix and recommend a typical mix as for:

(a) main concrete foundation blocks of a building on rock,
(b) cast-in-place reinforced concrete skeletal structure (slabs, beams, etc.),
(c) precast, prestressed shell roof units.

Indicate briefly the influence of time and environment on the compressive strength of concrete.

(C.E.I. Part 2)

16. Describe and explain four of the commoner forms of chemical attack on Portland cement concrete and indicate what precautions may be taken to minimise the effects of the attack.

(P.S.B. B.Sc. Bldg.)

17. What are "nominal mixes" and "standard mixes" in concrete technology ?
Outline the steps to be taken in designing a concrete mix for a particular situation of your choice.

What are the additional allowances to be taken into account in designing a concrete mix containing an air-entraining agent?

(S.O.E. Grad. Exam. C.E.)

18. Discuss, with reference to choice of materials, precautions and techniques involved in the following situations:

(*a*) cold-weather concreting,

(*b*) concreting in hot tropical conditions.

(S.O.E. Grad. Exam. C.E.)

FURTHER READING

Neville, A. M., *Properties of Concrete*, Pitman, 1973.

Lea, F. M., *Chemistry of Cement and Concrete*, Edward Arnold, 1970.

Orchard, D. F., *Concrete Technology Vols 1 and 2*, Applied Science Publishers Ltd., 1973.

Short, A. and Kinniburgh, W., *Lightweight Concrete*, C. R. Books, 1968.

Robson, T. D., *High Alumina Cements and Concretes*, Wiley, New York, 1962.

CHAPTER V

CLAY AND CLAY PRODUCTS

INTRODUCTION

Clay has been and is still used for building, pottery and modelling purposes. It is not easy to define clay precisely, as its definition varies according to different fields (*e.g.* geology, mineralogy, chemistry, soil science, civil engineering or ceramic technology). It can be regarded by the layman as a kind of natural earth which becomes plastic and mouldable when mixed with water, and becomes hard on drying and firing. It is characterised by its fineness (particle size about 2 microns) and is derived from the weathering and decomposition of igneous rocks or as a result of hydrothermal process which is associated with volcanic activity. Consequently it is composed mainly of silica (SiO_2), alumina (Al_2O_3) and water and is often contaminated with an appreciable amount of iron, alkalis and alkaline earths.

CLASSIFICATION OF CLAYS

Clay often contains groups of crystalline substances known as clay minerals such as quartz, feldspar, mica, etc. For convenience, clay minerals may be classified into two main groups:

1. Kaolinite group. Kaolinite may be regarded as a hydrous alumino silicate which can be represented by the general chemical formula:

$$Al_2(Si_2O_5)(OH)_4 \quad \text{or} \quad Al_2O_3 . 2SiO_2 . 2H_2O$$

Kaolinite occurs in soils of humid-temperate and humid-tropical regions.

Kaolinites are layer structures of silica tetrahedron (S) linked to gibbsite (G) which is hydrated aluminium oxide:

$$Al_2O_3 . 3H_2O \quad \text{or} \quad 2Al(OH)_3 \quad (\textit{see} \text{ Fig. 8}).$$

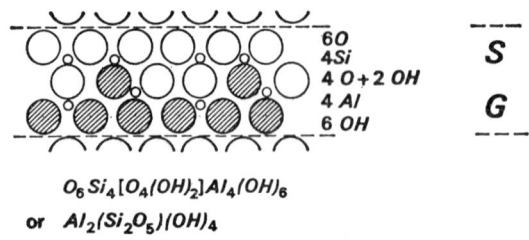

$$O_6 Si_4 [O_4(OH)_2] Al_4(OH)_6$$

or $Al_2(Si_2O_5)(OH)_4$

FIG. 8.—*Kaolinite*.

2. Pyrophyllite group. This consists of two layers of silica tetrahedra (S) separated by and linked to a gibbsite layer (G). The general chemical formula may be written as:

$$Al_2(Si_2O_5)_2(OH)_2 \quad \text{or} \quad Al_2O_3.4SiO_2.H_2O$$

and the structure as shown in Fig. 9.

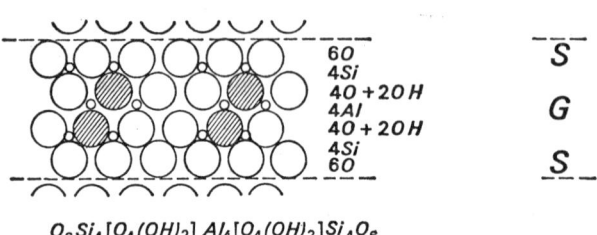

$$O_6 Si_4 [O_4(OH)_2] Al_4 [O_4(OH)_2] Si_4 O_6$$

or $Al_2 (Si_2 O_5)_2 (OH)_2$

FIG. 9.—*Pyrophyllite*.

Derived from the pyrophyllite group are:

Montmorillonite group, in which some of the Al^{3+} ions in the gibbsite layer are replaced by Mg^{2+}, Na^+, K^+ or Ca^{2+} ions. Montmorillonite occurs in soils where rainfall is low, *e.g.* in deserts and prairies. It is the dominant clay mineral in bentonite (a modified form of volcanic ash) and it is more colloidal in nature than kaolinites.

Illite group, in which some of the Si^{4+} ions in the silica tetrahedron are replaced by Al^{3+} ions together with Na^+ or

K^+ ions. Illites are mica-like clay minerals, also known as hydrous micas. They are the chief constituent of many shales.

PROPERTIES OF CLAY

Clay properties are influenced by such factors as chemical composition and structure, particle size and shape.

3. Colloidal properties. Colloidal systems consist of a finely-divided disperse phase more or less uniformly distributed throughout a continuous phase known as a dispersive medium, each being insoluble in the other; each may be solid, liquid or gas but not gas and gas. Some examples are given in Table XX.

TABLE XX: DISPERSE SYSTEMS

Dispersed particles	Dispersive medium	Examples
solid	solid	some alloys, glasses (ruby)
liquid	solid	solid emulsions, some gels
solid	liquid	sols, colloidal solutions
liquid	liquid	emulsions
gaseous	liquid	gas dispersions, foams
solid	gaseous	dust, smoke, solid aerosol
liquid	gaseous	fog, mist, liquid aerosol

The size of colloidal particles varies between 0.0005 μm and 0.2 μm. Two main types of colloidal systems are recognised:

Lyophilic (*e.g.* glue, starch, gelatin) which forms a very stable dispersion. The term "hydrophilic" is used if the dispersive medium is water. Generally, lyophilic (indicating affinity for liquid) colloids or sols are reversible, *i.e.* they can be re-dissolved after precipitation.

Lyophobic (*e.g.* Prussian blue, arsenious sulphide) which does not readily form a stable dispersion. The term "hydrophobic" is used if the dispersive medium is water. Lyophobic (indicating dislike for liquid) colloids or sols are generally irreversible, *i.e.* they cannot be redissolved readily.

Clay particles are colloidal in size and can be dispersed in excess water to form a colloidal solution. Clays are generally hydrophobic but some (particularly the montmorillonite group) possess hydrophilic properties. The colloidal suspension may set to a gel on standing but liquefy on shaking and these processes can be repeated indefinitely. The suspension is said to be *thixotropic*.

All colloidal systems exhibit "Tyndall effect" (the phenomenon of light scattering), "Brownian movement" (the random motion of colloidal particles in a fluid) and also "electrophoresis" (the movement of colloidal particles through a liquid under the influence of an electric field) thus showing that colloid particles carry an electric charge. Clay minerals are negatively charged (anions) whereas hydrogen and metallic ions are positively charged (cations). The charge possessed by

TABLE XXI: SIZE AND SHAPE OF CLAY MINERAL PARTICLES

Kaolinite	Hexagonal plates, size usually 0·1–3 μm
Montmorillonite	Very poorly defined hexagonal plates, much less than 1 μm
Illites	Poorly defined, thin hexagonal flakes, about 1 μm in size

the clay minerals may originate as a result of unbalanced ionic substitutions in the crystal structure or absorption of ions at active sites often associated with broken bonds. Table XXI gives the particle size and shapes of three main groups of clay minerals.

4. Ion-exchange properties. Ion-exchange is a chemical process involving the reversible interchange of ions (charged particles) at a phase boundary, as between a solution and a particular solid material such as clay. The ability of clays to exchange ions (whether cations or anions) is a measure of the *ion-exchange capacity*, which is usually expressed in terms of milliequivalent per 100 grammes of dried solid.

Cation-exchange. Inter-layer Na^+ ions of swollen montmorillonites, for example, can be exchanged by Ca^{2+} ions if hard water is passed through the hydrated clays — Na^+ ions passing into solution. This cation-exchange property is typical of clays, and it can be represented by the simple equation:

$$X . \text{Clay} + Y^+ \rightleftharpoons Y . \text{Clay} + X^+$$

Table XXII gives typical values for cation-exchange capacity of some clay minerals.

TABLE XXII: CATION EXCHANGE CAPACITY
(IN MILLIEQUIVALENT/100 GM)

Kaolinite	3–15
Montmorillonite	80–150
Illite	10–40
Vermiculite	100–150

Anion-exchange. It is possible, but less likely, for clays to exhibit anion-exchange properties. A simple equation can be represented as follows:

$$\text{Clay} . B + A^- \rightleftharpoons \text{Clay} . A + B^-$$

Table XXIII gives typical values for anion-exchange capacity of some clay minerals.

TABLE XXIII: ANION-EXCHANGE CAPACITY
(MILLIEQUIVALENT/100 GM)

Kaolinite (colloidal)	20·2
Montmorillonite	31
Vermiculite	4

5. Plasticity. Clays possess the unique and characteristic property of becoming plastic when mixed with a limited amount of water. Plasticity, for practical purposes, may be regarded as rather similar to workability.

Plasticity may be defined as the property of a material which allows it to be deformed under the action of stress without rupturing and to retain the shape produced after the removal of stress.

Various methods of assessing plasticity have been suggested.

One method involves the determination of the water of plasticity (*see* Table XXIV), *i.e.* the amount of water required to give optimum plasticity or consistency as judged subjectively by the operator. Another method makes use of the penetrometer principle—by determining the amount of penetration of a standard needle or plunger into a plastic mass of clay when subjected to a given load or rate of loading.

TABLE XXIV: WATER OF PLASTICITY
(% OF DRY WEIGHT)

Kaolinite	8·9–56·3
Montmorillonite	83–250
Illite	17–38·5

Plasticity is dependent on the moisture adsorbed—the greater the moisture adsorbed the greater the plasticity. Other factors influencing plasticity include particle size, particle shape, nature of surface, applied pressure and presence of electrolytes.

6. Dry strength is a measure of the resistance which a bar or rod of clay (after drying, usually at 105 °C) offers to a load applied at right angles to its length. It is expressed in terms of the modulus of rupture (*see* Table XXV).

TABLE XXV: DRY STRENGTH OF CLAY MINERALS
(N/mm^2)

Kaolinite	0·07–4·84
Illite	1·48–7·40
Montmorillonite	1·89–5·70

Dry strength is mainly governed by the nature and size of the clay minerals. Other influencing factors include moisture content of dry clay and plasticity of the mix.

7. Drying and firing shrinkage. After the moulding of clay products, the added water is removed by drying (usually at 105 °C). This drying process is accompanied by a contraction equal to the volume of water (pore water + adsorbed water) driven off. The amount of contraction is a measure of *drying shrinkage*.

However, if heating is carried out at a much higher temperature (say, at 1000 °C or above), further shrinkage, known as *firing shrinkage*, takes place, when fusion or vitrification occurs. Table XXVI gives typical values of drying and firing shrinkages of some clay minerals.

TABLE XXVI: DRYING AND FIRING SHRINKAGES OF
CLAY MINERALS

Mineral	Linear drying shrinkage (%)	Linear firing shrinkage (%)
Kaolinite	3–10	2–17
Illite	4–11	9–15
Montmorillonite	12–23	±20

Excessive drying shrinkage, if not uniform within the product, often leads to cracking and warping. It is a severe limitation to the use of highly plastic clays.

TYPES OF CLAY

8. China clays or kaolins. Residual or sedimentary. They are free from iron (hence can be fired to a white pure colour) and free from alkalis (hence highly refractory). Plasticity varies from poor to fairly good. Washed kaolins fuse at 1750–1780 °C.

Obtained as a result of weathering of granite (containing alkali feldspar):

$$K_2O.Al_2O_3.6SiO_2 + CO_2 + 2H_2O$$

alkali feldspar   carbon   water
                    dioxide

$$\longrightarrow Al_2O_3.2SiO_2.2H_2O + 4SiO_2 + K_2CO_3$$

China clay

9. Ball clays. Sedimentary, ball clays fire to a white or cream colour with excellent plasticity and dry strength. Very finely divided. Ball clays fuse at 1650–1780 °C (highly refractory).

10. Fire clays. Sedimentary, not white burning (colour due to contamination of iron oxide) but free from fluxes such as iron, alkaline earths, alkalis and excess silica. Highly refractory, fire clays do not fuse below 1600 °C.

11. Brick clays. Very impure form of clays which can be used without further treatment and fired at a low temperature (1160–1300 °C) to give satisfactory products of acceptable colours and textures.

12. Bentonites. Derived from glassy volcanic ash. Colour varies from white to light blue/green. With water, these swell more than any other dried clay to a soft gel which is strongly thixotropic. Sodium bentonites exhibit larger volume changes than calcium bentonites.

Used in the construction of dams or reservoirs for preventing water seepage and also for increasing the plasticity of poor plastic clays.

13. Fuller's Earths. Residual or sedimentary. Consist mainly of hydrogen bentonites. Do not swell in water and are acidic in reaction. Useful for decolourising oils.

CLAY PRODUCTS

14. Manufacture of clay products (structural). Generally, this involves four stages:

(a) *Preparation*, including winning and pre-treatment.

(b) *Shaping*, moulding and pressing (by hand or machine).

(c) *Drying*. This is essential to remove water before firing, otherwise steam generated may disrupt the material, sometimes explosively. Natural drying or artificial drying in a kiln can be used.

(d) *Firing*, up to a temperature of 1000 °C or more. Remaining water is removed and the clay body changes into a hard, rigid material.

15. Effects of firing. Four stages can be recognised:

(a) *Drying* (up to about 100° C). Water of plasticity

(shrinkage water) is removed and the clay product becomes rigid but brittle.

(b) *Dehydration* (between 100 °C and 700 °C). The clay minerals lose their water of crystallisation (kaolinite loses its water of crystallisation between 450 ° and 510 °C).

(c) *Oxidation* (between 550 °C and 900 °C). Iron compounds are oxidised to ferric oxides (Fe_2O_3) and all carbonaceous impurities are burnt out before the temperature reaches 800 °C.

(d) *Vitrification* (950 °C and above). Crystallisation of mullite ($Al_6Si_2O_{13}$) begins, the mullite crystals grow as the temperature is increased and the "glassy" phase contracts, causing a severe shrinkage.

16. Types of clay products.

(a) *Heavy or structural clay products.*

   (i) Building bricks (BSS 187, 1758, 2973, 3056, 3679, 3921, 4729).
   (ii) Structural tiles (BSS 402, 1281, 1286, 3679).
   (iii) Clay blocks (BS 3921).
   (iv) Sewer pipes (BSS 65 and 540, 539, 1143, 1196).
   (v) Porcelain and porcelain enamels (vitreous enamels) (BSS 914, 3830, 3831).

(b) *Cements.* E.g. various types of Portland cements.

(c) *Lightweight aggregates* (BSS 3681, 3797). E.g. expanded clay, expanded shale, exfoliated vermiculite.

(d) *Refractories* (BSS 1758, 2973, 3056). These are materials which can withstand temperatures not less than 1500 °C without softening or deformation. E.g. fireclay and bauxite.

PROGRESS TEST 5

1. What is clay ? (p. 77)
2. How are clay minerals classified ? (1–2)
3. What are the main properties of clays ? (3–7)
4. Outline the changes which take place during the firing of clays. (7, 15)
5. Compare the nature and properties of the different types of clays. (8–13)

6. Name the different types of clay products available commercially. (16)

1. A modern block of flats having brickwork walls and solid concrete floors has large patches of dampness affecting bedroom walls.

(a) Discuss the likely cause of the dampness.

(b) Suggest suitable remedial measures for curing the problem.
(I.O.B. Assoc. Part 1 Specimen)

2. (a) Describe the stages in the firing of a clay brick, indicating the significance of organic matter, iron salts, silica and alkali metal salts present in the clay.

(b) Explain the terms DRYING SHRINKAGE and FIRING SHRINKAGE and discuss the relationships between firing temperature, fire shrinkage, porosity and frost resistance for a particular clay product.

(P.S.B. H.N.D. Bldg.)

3. (a) Describe three methods of manufacturing machine-made clay bricks, mentioning the appearance and properties produced by each method.

(b) Explain the type and causes of failure which may occur when a rendering of cement and sand is applied to the exposed exterior face of a wall built in Fletton brickwork.

(c) Indicate three types of pointing to brickwork and explain the reasons for pointing.

(P.S.B. B.Sc. E.M. & B.E.)

4. Give the provisions of BS 3921 Part 2 concerning the following physical properties of clay bricks:

(i) strength,
(ii) absorption,
(iii) efflorescence.

Discuss to what extent these properties are related to the probable durability of bricks.

Describe how the rate of water absorption of bricks affects the operation of brick laying.

(P.S.B. B.Sc. S.E.)

FURTHER READING

Chandler, M., *Ceramics in the Modern World*, Aldus Books, London, 1967.
B.D.A. Book Series, Brick Development Association Ltd.
Keeling, P. S., *The Geology and Mineralogy of Brick Clays*, 1963.

Goodson, F. J., *Clay Preparation and Shaping*, 1962.
Ford, R. W., *The Drying of Bricks*, 1967.
Rowden, E., *The Firing of Bricks*, 1964.
West, H. W. H., *The Layout of Brickworks*, 1963.
Clews, F. H., *Heavy Clay Technology*, British Ceramic Society, 1955.

CHAPTER VI

GLASS

INTRODUCTION

Glass is one of the oldest and most versatile materials known to man. It occurs naturally as a volcanic glass known as *Obsidian*. Artificial glass was made at least 4000 years ago.

The term "glass" in its widest sense is used to describe a particular state of matter, known as the glassy or vitreous state, which is obtained when a liquid cools without crystallisation taking place. The product has the apparent physical properties of a solid which is characterised by its brittleness, hardness, transparency and chemical inertness.

MANUFACTURE

The main product used in construction is the flat glass product. The various steps in its manufacture are shown schematically in Fig. 10.

RAW MATERIALS

1. Glass-formers. Sand (Silica SiO_2) is the main glass-former. It is highly resistant to chemical attack and it has an extremely high melting point.

2. Modifiers. These are materials added in order to lower melting and working temperatures by decreasing viscosities (*e.g. Na_2O, K_2O, B_2O_3*—known as *fluxes*), to improve chemical durability and/or to prevent crystallisation (*e.g. CaO, MgO, Al_2O_3*—known as *stabilisers*).

3. Melting and refining agents. These are materials added in small amounts in order to remove tiny gas bubbles from the glass (refining process). Common examples are sodium

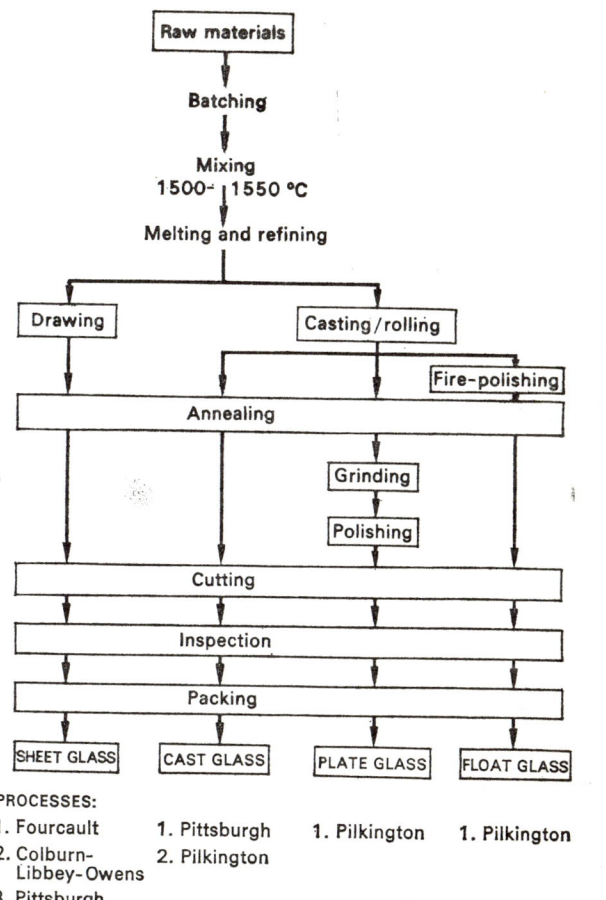

Fig. 10.—*Manufacture of flat glass product.*

sulphate, sodium nitrate, sodium chloride, arsenious oxide, calcium fluoride and carbon.

4. Colouring and decolourising agents. These are materials which are added in small amounts either to produce a desired colour or to eliminate an undesired colour. Colouring agents, together with the colour produced, include:

chromic oxide (green)
cobalt oxide (blue)
cuprous oxide (red)
ferrous oxide (blue)
ferric oxide (brown)
selenium (pink)
uranium oxide (yellow)

Decolourising agents are:

selenium
cobalt oxide
neodymium oxide

5. Opalising agents. These are materials which are added to impart an opalescent appearance to the finished glass. Examples are calcium fluoride (used for opacifying soda-lime glasses), calcium phosphate (used for opacifying borosilicate glasses).

Typical batch composition for flat glass:

Raw materials	Batch weight (kg)	Role
Sand	1000	Main constituent and glass-former (SiO_2–quartz)
Soda ash (Na_2CO_3)	310	Fluxing agent (Na_2O)—to lower the melting temperature of SiO_2
Limestone ($CaCO_3$)	55	CaO stabilising oxide to reduce solubility of sodium silicates
Dolomite ($MgCO_3$)	240	MgO stabilising oxide
Feldspar ($K_2O . Al_2O_3 . 6SiO_2$)	60	Al_2O_3 stabilising oxide—to improve chemical durability
Sodium sulphate (Na_2SO_4)	15	Refining agent—to remove gas bubbles from the glass
Cullet	400	Broken glass—to improve the meltability of the batch

TYPES OF GLASS

Glass may be regarded as complex form of silicate. For convenience, it can be classified, according to chemical composition, into the following types.

6. Silica glass (fused silica). This belongs to one of the most important single-oxide glasses. It is characterised by its highly cross-linked, three-dimensional bonding, and consequently it has high melting and working temperatures, low coefficient of thermal expansion and high chemical resistance. This type of glass is suitable for use as laboratory ware. A less expensive type contains 96 per cent silica, known as 96 per cent Silica glass.

7. Soda–lime–silica glasses. The presence of soda (Na_2O), which acts as a fluxing agent, reduces the melting and working temperatures, while the presence of lime (CaO), which acts as a stabilising oxide, reduces the solubility of sodium silicates. This type of glass has wide application in the manufacture of glass containers, flat and plate glass, domestic ware, electric lamp bulbs, etc.

8. Borosilicate glasses. The presence of boric oxide (B_2O_3), which acts as a flux as well as a glass-former, gives borosilicate glasses a low coefficient of thermal expansion and therefore a better thermal shock resistance. In addition, the presence of alumina (Al_2O_3) improves chemical durability and resistance to devitrification (crystallisation). "Pyrex" glass belongs to this type, which is greatly used in laboratory and medical ware.

9. Aluminosilicate glasses. The "hardening" effect of this type of glass is due to the presence of Al_2O_3 which also improves chemical durability and resistance to devitrification. One such type, known as E-glass (lime aluminoborosilicate), is used in the manufacture of glass fibres.

10. Lead glasses. The presence of lead oxide (PbO) acts as a flux and a modifier, hence lead glasses have usually low melting

TABLE XXVII: PROPERTIES

Type of glass	Approximate formula and structure	Typical composition (%)						
		SiO_2	Al_2O_3	B_2O_3	Na_2O	K_2O	MgO	CaO
Silica	SiO_2 O—Si—O—Si— —Si— O O—Si—O—Si—	99·9						(H_2O) 0·1
Soda–lime–silica	$Na_2O.CaO.5SiO_2$ $\ominus O$—Si—O—Si—O—Si—$O\ominus$ $Ca\oplus\oplus$ $Na\oplus$ $Na\oplus$ $\ominus O$—Si—O—Si—$O\ominus$	(1) 73 (2) 72·5	1 1·3		17 15·9	 0·3	4 3·0	5 6·5
Borosilicate	$\frac{1}{2}B_2O_3.4SiO_2$ O—Si—O—B< —Si— O—Si—O—Si—	(1) 81 (2) 75·7 (3) 74·7	2 5·1 5·6	13 6·9 9·6	4 6·2 6·4	 1·2 0·5		 1·3 0·9
Alumino-silicate	$Na_2O.Al_2O_3.3SiO_2$ O—Si—O—Al—$\ominus Na\oplus$ —Si— O O—Si—O—Al—$\ominus Na\oplus$	(1) 62 (2) 57	17 15	5 5	1		7 7	8 10 6 (BaO)
Lead	$2PbO.3SiO_2$ O—Pb—O—Si— —Si— O O—Pb—O—Si—	(1) 56 (2) 3 (3) 5	2 11 3	 11 10	4	9		(PbO) 29 75 82

and working temperatures but high refractive indices and densities. High lead type (Table XXVII composition (3)) is suitable for use in radiation shielding.

The properties of glasses are governed by various factors, the main one being their chemical structure and composition. Some important properties are summarised in Table XXVII.

OF GLASSES

Chemical durability[1] weathering water acid			Thermal expansion ($\times 10^{-7}$ K^{-1})	Young's modulus (E) (kN/mm^2)	Refractive index (n_D)	Specific gravity
1	1	1	5·5	72	1·459	2·20
3	2	2	92 93	69	1·512	2·47
1	1	1	33 50	63 71	1·474 1·49	2·23 2·39 2·36
1	1	3	42	87	1·530	2·52
1	1	3	46	86	1·547	2·64
2	2	2	89	59	1·560	3·05
1	1	4	84	55		5·42
3	1	4	104	51	1·97	6·22

Chemical durability in decreasing order 1, 2, 3, 4.

GENERAL PROPERTIES OF GLASS

11. Chemical properties. Silicate glasses are fairly chemically resistant under normal conditions. They are not readily attacked by water, dilute acids, alkalis and salts but are affected by such chemicals as lime, concentrated solutions of caustic soda ($NaOH$), hydrofluoric acid (HF) and fluorides. The degree of chemical resistance is, however, dependent on the chemical composition of glass (see Table XXVII under chemical durability). Phosphate glasses (used as optical glasses) based on phosphorus pentoxide (P_2O_5) are highly resistant to hydrofluoric acid, and by adding some iron oxide to the glass composition, become heat-absorbing.

12. Mechanical properties.

(a) Specific gravity. Glass is generally light and has a specific gravity of around 2·5 which is comparable to aluminium (specific gravity 2·7). Lead glasses, due to the heavier constituent (lead oxide, PbO), may have a specific gravity above 6·0 (Table XXVII).

(b) Strength. This is perhaps the most anomalous property of glass, and there does not seem to be a reasonable correlation between strength values and glass composition. At ordinary temperatures, glass is not plastic—i.e. it breaks as soon as its elastic limit is exceeded. Young's modulus of elasticity ranges from 50 to 87 kN/mm² (approx.), tensile strength is around 0·1 kN/mm², and compressive strength around 1 kN/mm², whereas the strength of glass fibres is very high, depending on diameter. Tensile strength of glass fibre (diameter 8–10 μm) = 3 kN/mm², and increases with the decrease in diameter of the fibre.

However, glass can be strengthened by:

(i) Thermal toughening (tempering of glass). This involves cooling the surfaces of glass rapidly from annealing temperatures. The disadvantage of this process is that when glass eventually breaks severe fracturing may occur causing a safety problem, since the toughened glass is already in a state of strain.

(ii) Chemical toughening. This makes use of the ion-exchange principle—by replacing the smaller sodium ion (Na^+) in the glass surface by a larger potassium ion (K^+) the surface of glass becomes "compressed" and consequently toughened.

(c) Hardness. Hardness (surface) is associated with strength. Hardness of glass is 5·4–5·8 on the Moh's scale (diamond has hardness 10). Surface hardness can be increased by the thermal toughening process to the value 9, which is as hard as corundum (see Appendix II).

13. Optical properties.

(*a*) Refractive index (*n*) and dispersion (*d*). In general, refractive index (*n*) of glass can be defined as:

$$n = \frac{\text{Sin } i}{\text{Sin } r} = \frac{\text{velocity of light in vacuo}}{\text{velocity of light in glass}}$$

where *i* and *r* are the angles of incidence and refraction respectively. The refractive index, therefore, defines the concept of the deviation of light when passing from one medium (such as air) to another medium (such as glass). It is dependent on the wavelength of the incident light. The mean values of the refractive index (n_D) of the various types of glass are given in Table XXVII, where n_D is the refractive index at sodium *D*, wavelength 589·58 nm. *Dispersion* of light is the separation of composite white light into its constituent colours (which are wavelength-dependent). Dispersive power (*ω*) is defined as:

$$\omega = \frac{n_F - n_c}{n_D - 1},$$

where n_c, n_F are the refractive indices at Hydrogen *C* and *F* lines, wavelengths 656·28 nm and 486·13 nm respectively.

The properties of optical glasses (used in optical instruments, lenses, etc.) are defined by *n* and *d*, both of which are dependent upon glass composition.

(*b*) Reflectance (*R*), Absorptance (*A*) and Transmittance (*T*). These govern the transparency which is one of the prime properties of glass.

A beam of light incident at an angle on a block of glass undergoes the phenomena of reflection, refraction and transmission as shown in Fig. 11. The proportion of light (luminous flux) that is reflected, absorbed or transmitted is conveniently expressed in terms of a factor respectively known as Reflectance (*R*), Absorptance (*A*) and Transmittance (*T*).

Reflectance (*R*) is related to the refractive index (*n*) according to the Fresnel formula:

$$R = \left[\frac{n-1}{n+1}\right]^2;$$

assuming a common glass where *n* = 1·5, *R* becomes 0·04

or 4 per cent (for each surface). From Fig. 11, Transmittance (*T*) will be less than 92 per cent considering that part of the light is always absorbed by the glass material. *T* varies according to the optical properties of glass and the wavelength of the incident light. Absorptance (*A*) also varies

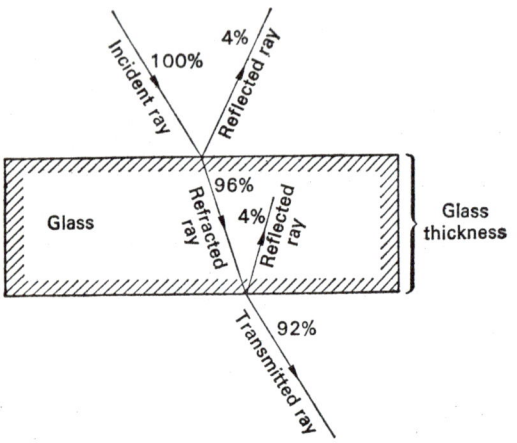

Fig. 11.—*Light path through a block of glass.*

with the wavelength of light. *R*, *A* and *T* all vary with the angle of incidence of the beam of the light, especially after 40°.

14. Thermal properties.

(*a*) *Specific heat capacity* (*C*). This is defined as the quantity of heat (measured in joules (J)) required to raise the temperature of unit mass (kg) of a substance by one degree rise in temperature (K).

For glass, $C = 0.84$ kJ/kg K, which is about one-fifth that of water.

(*b*) Thermal conductivity (*k*). This is the ability to transfer heat by conduction and is defined by the equation:

$$Q = \frac{kA(\theta_1 - \theta_2)}{X}$$

where Q = quantity of heat conducted through the material per second (W)

A = area of surface (m^2)

X = thickness (m)

θ_1, θ_2 = temperature of opposing surfaces (K).

Ordinary glass is a poor conductor of heat,

k = 0·8–1·05 W/mK, which is comparable to that of building bricks, concrete, water.

(c) U-value (or coefficient of thermal transmittance). U-value can be calculated from the relation:

$$U = \frac{1}{\dfrac{1}{f_1} + \Sigma \dfrac{l}{k} + \dfrac{1}{f_2}},$$

where l = thickness of material (m)

k = thermal conductivity (W/mK)

f_1 = *inside* film or surface conductance (= 8·14 W/m^{2}K normally)

and f_2 = *outside* film or surface conductance (= 21 W/m^{2}K normally).

Average U-value for glass is 5·82 W/m^{2}K (this varies very slightly with thickness).
Table XXVIII gives U-values for some building materials.

TABLE XXVIII: U-VALUES OF BUILDING MATERIALS

Materials	*U-value* (W/m^{2}K)
Glass: Single glazing	3·98–7·38
Double glazing with gaps:	
5 mm	2·78–4·09
6 mm	2·67–3·81
12 mm	2·39–3·36
19 mm	2·33–3·18
Glass blocks (80 mm)	2·5
Solid brick wall (105 mm) with dense plaster	
(16 mm)	3·0

(d) Thermal expansion. Like most materials, glass expands on heating. This is, however, true only up to a temperature known as the annealing point. Figure 12 shows a typical expansion curve of a glass.

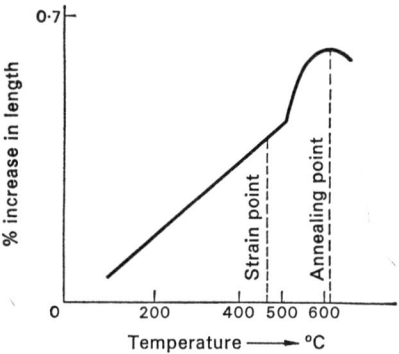

FIG. 12.—*Typical thermal expansion curve of a glass.*

Coefficient of linear expansion (α) of glass ranges from about 5×10^{-7} to over $100 \times 10^{-7} \, K^{-1}$ depending on type of glass (*see* Table XXVII). Differential thermal movement is often a problem when glazing in metal sashes (α for steel and aluminium $= 120 \times 10^{-7}$ and 240×10^{-7} K^{-1} respectively). Normal glazing-quality glass cracks on heating (thermal shock). Silica and borosilicate glasses are more resistant to thermal shock. So are toughened glasses.

(e) Effect of fire. Normally glasses are non-combustible but crack and melt due to the high temperatures encountered in a fire. Fire resistance, however, can be improved by increasing thickness and/or incorporating thermal conducting materials as in wired or copperlight glazing.

15. Acoustic properties. The sound insulation value of a structure is usually expressed in terms of transmission losses. It is a measure of the amount of sound intensity (in decibel unit—dB) reduced when passing through the structure.

Single glazing has a relatively low sound insulation value, dependent on the thickness of the glass. Double or triple glazing improves the sound insulation, which is mainly dependent on the air space in between the glass panes. Some sound insulation values are given in Table XXIX.

TABLE XXIX: SOUND INSULATION VALUES OF
BUILDING MATERIALS

Materials	Insulation value (dB)
Single sheet:	
Aluminium (0·6 mm thick)	16
Galvanised iron (0·7 mm thick)	32
Plywood (3-ply, 3 mm thick)	26
Fibreboard (12 mm thick)	22
Sheet glass (3 mm thick)	28
Plate glass (6 mm thick)	32
Double sheet:	
Sealed double window glass with	
200 mm of air space	40

16. Electrical properties. At room temperature glass can be considered as an electrical insulator, but at high temperatures it becomes a good conductor of electricity: the electric resistivity of glass can readily decrease from 10^{10} Ωm (at room temperature) to 10^{-1} Ωm (at 1400 °C) and can increase to 10^{17} Ωm if the surface is treated with a water-repellent agent such as silicone. Electric melting methods in glass furnaces make use of the high electrical conductivity of glass at high temperatures.

Some electrical data for ordinary glass at 20 °C are given:

Dielectric constant	6–7
Dielectric strength in air at 50 Hz	3×10^7–8×10^7 V/m
Resistivity	10^9–10^{10} Ωm

GLASS FIBRE PRODUCTS

The important products are the glass "wools" and the glass-fibre reinforcements. These are mainly used for insulation purposes (thermal, sound and electrical) and as reinforcements in plastic products (Glass Reinforced Plastics GRP), in plasters (Glass Reinforced Gypsum plasters GRG) and in cements (Glass Reinforced Cements GRC).

The types of glasses used are mainly the A-glass (Soda–lime–silica glass), the E-glass (alumino–borosilicate glass of low alkali lime) and the AR-glass (alkali resistant glass containing zirconia).

Some important composites are listed in Table XXX.

TABLE XXX: GLASS-FIBRE PRODUCTS OR COMPOSITES

Composites	Matrix	Glass fibre	Remarks
GRP	Polymer resin	A- or E-glass	See IX, 12, 17, 19, 23, 24, 33.
GRG	Gypsum plaster	E-glass	Improved mechanical strength and fire resistance. Not suitable for exterior use since strength is affected on wetting.
GRC	OPC	AR-glass	Improved mechanical strength and fire resistance. A- or E-glass not suitable for use with OPC, due to alkali attack.

For details consult BRE Current Papers 67/68, 33/69, 40/69, 40/70, 12/71, 54/74, 79/74, 26/75, 65/75, 94/75, 38/76 (H.M.S.O.)

VITREOUS COATINGS

Vitreous coatings (glazes) can be applied on:

(a) Ceramic surfaces (glazing of ceramic surfaces).
(b) Metal surfaces (vitreous or porcelain enamelling).

17. Glazing of ceramic surfaces (BSS 1344, 1359, 3402, 3831). The glazing compounds (SiO_2 + mixes of the various oxides such as B_2O_3, Al_2O_3, BeO, K_2O, Li_2O, CaO, MgO, SrO, BaO,

ZnO, PbO, SnO_2, ZrO_2) are heated to melting in a bath. By dipping the ceramic object into the glazed bath, a vitreous coating is left on the ceramic surface. The thickness of the coating can be controlled by the duration of immersion in the bath, by the rate of withdrawal from the bath and by the porous nature of the ceramic surface.

18. Vitreous enamelling (BSS 1344, 1358, 3830). Metal surfaces (such as steel, cast iron, aluminium, copper) can be coated with glass. These surfaces should be thoroughly cleaned (*e.g.* sandblasting, etc.) prior to treatment. Two methods are available:

(*a*) *Wet process*. The glazing compound consisting of silica and mixed calcium, nickel and aluminium salts is made into a slurry with water. The slurry is then painted on the metal surface at room temperature. Heat is gently applied to the surface so that water is driven off by evaporation and eventually raised slowly to red heat until the compound melts and fuses. A chemical reaction takes place resulting in a vitreous coating being bonded to the surface through the formation of silicate.

(*b*) *Dry process*. The metal is first heated and the glazing compound carefully and uniformly applied over the red-hot surface. By reheating, the compound fuses and becomes bonded to the metal surface. The thickness of the coating obtained varies from 0·8 to 1·6 mm. The dry process is more expensive but more efficient than the wet process.

APPLICATIONS OF GLASS

19. Packaging. Glass containers and bottles are very widely used. They are sufficiently strong, chemically inert and resistant to thermal shock. Durability and resistance to abrasion can be improved by a process known as "Titanising" in which a thin transparent coating of titanium oxide is deposited on the surface of glass. Mechanical strength can be improved by a "toughening" process and by judicious design (BS 4602—The use of metric units in specifications for glass containers and finishes).

20. Applications in construction and engineering.

(*a*) Windows and doors in buildings use various forms of glass.

> (*i*) Transparent glasses (clear plate and sheet glass).
> (*ii*) Translucent glasses (including wired, ribbed and patterned glass).
> (*iii*) Toughened glasses (*e.g.* Armourplate, Armourclad, Armourglass—manufactured by Pilkington Bros. Ltd.—BSS 857, 3925).
> (*iv*) Decorative glasses (such as glass mosaic, fused glass, slab glass used in concrete, stained-glass windows, *e.g.* in churches and cathedrals).

(*b*) Glass blocks (for use in walls, staircase lights and panels)

> (*i*) Translucent glass blocks.
> (*ii*) Preformed hollow blocks, (BS 1207, CP 122).

(*c*) Double glazing of windows, etc. (BS 952, CPP 145, 152)

> (*i*) Prevents heat loss; hence saving of fuel costs.
> (*ii*) Prevents condensation and misting.
> (*iii*) Stops draughts.

(*d*) Glass claddings in buildings. Improves appearance, durability, fire resistance and insulation.

(*e*) Foam glass. Used for thermal insulation in refrigerators, building walls and roofs.

(*f*) Glass fibres (BSS 3396, 3496, 3691, 3749, 3779, 3953, 4045, 4584: Part 7).

> (*i*) Thermal and electrical insulations, and sound absorption.
> (*ii*) Reinforcing material for plastics (Glass-reinforced plastics), cement (Glass-reinforced cement).
> (*iii*) Gypsum plaster (Glass-reinforced gypsum plaster).

(*g*) One-way mirrors. Include venetian mirrors and transparent mirrors.

(*h*) Electrical uses (BS 4145). Light bulbs, fluorescent tubes, neon lighting, street lighting, in ultrasonics and electronics.

(*i*) Domestic and scientific glass wares (BSS 1751, 1797, 3193, 3517). Including "Pyrex" glasses.

(*j*) Glass ceramics. Missile nose cones, domestic oven

wares (Pyroceram in U.S. and Pyrosil in U.K.).

(k) Fibre optics.

(l) Radiation shielding (BS 4031). Lead glasses (high lead content).

PROGRESS TEST 6

1. What are the ingredients used in glass manufacture? **(1–5)**
2. How is glass classified chemically? **(6–10)**
3. Give an account of the general properties of glass. **(11–16)**
4. Mention some of the important glass-fibre products commonly used. **(pp. 99–100)**
5. Describe the various methods of applying vitreous coatings to ceramic and metal surfaces. **(17–18)**
6. What are the main building applications of glass? **(20)**

EXAMINATION QUESTIONS

1. Describe the molecular structure of crystalline quartz and the changes in structure involved in the formation of soda glass. Compare and contrast the atomic bonding in silica and in a paraffin, and show how these bonding types are related to the gross properties of the materials.

(B.S.B. B.Sc. S.E.)

2. (a) "Glass is a supercooled liquid." Discuss the truth (or otherwise) of this statement.
   (b) How would you compare float glass and plate glass from the standpoint of:
       (i) production,
       (ii) cost,
       (iii) light refraction,
       (iv) thickness available.
   (c) Compare the wet and dry processes of vitreous enamelling of metal surfaces.

(P.S.B. B.Sc. S.E.)

3. (a) Describe the production, characteristics and uses of glass fibre.
   (b) What is meant by the devitrification temperature of glass?
   (c) Discuss the nature and properties of:
       (i) tempered glass,
       (ii) safety glass,
       (iii) heat-resisting glass.

(P.S.B. B.Sc. S.E.)

FURTHER READING

Maloney, F. J. T., *Glass in the Modern World*, Aldus Books Ltd., 1967.
Persson, R., *Flat Glass Technology*, Butterworth, 1969.

CHAPTER VII

TIMBER

INTRODUCTION

Timber is a very versatile building material which has been used by man since his most primitive days. Because it is a living material, it is highly complex in nature and therefore structurally it is less well known than steel or concrete. However, better understanding and knowledge of laminating, coupled with the wider availability of synthetic glues and adhesives, has overcome many of the limitations in its use as an engineering material.

SOURCE OF TIMBER

1. Structure of the tree. This consists of three parts:

(*a*) The *root system*, which

(*i*) absorbs water and mineral salts (such as nitrate and phosphates) from the soil, and

(*ii*) anchors the tree firmly on the ground so that it can resist the overturning effect of the wind.

(*b*) The *trunk or stem*, which

(*i*) provides mechanical strength and supports the crown at such a height above its surroundings to ensure sufficient access of sunlight and air,

(*ii*) conveys food from the leaves to the soil, and vice versa.

(*c*) The *crown or the branch system*, which spreads out the leaves to form a catchment area large enough for the manufacture of food from sunlight and air (photosynthesis).

COMPOSITION OF WOOD

Apart from water which is present up to 150 per cent of dry weight (oven-dried at 100–105 °C) in the growing tree, the constituents of wood are mostly organic matter.

2. Cellulose: 45–60 per cent of dry weight of wood. This is the main constituent of cell-wall carbohydrates built up of glucose units. On hydrolysis with acid it forms sugars which give rise to alcohols on fermentation. It is an important raw material in wood in the pulp and paper, rayon and explosive industries.

3. Hemicellulose: 15–25 per cent of dry weight of wood. This is also a constituent of the cell-wall carbohydrates, the length of chain molecule being shorter than in cellulose. More easily hydrolysed than cellulose giving sugars.

Cellulose and hemicellulose are represented by the same simplified formula $(C_6H_{10}O_5)_n$, where n is a very large number.

4. Lignin: 25–35 per cent of dry weight of wood. A non-carbohydrate component forming the lining of, and strengthening, the cell wall. It can be extracted by means of caustic soda or sodium bisulphite. One of the by-products, lignosulphonate, is used as concrete admixture.

5. Inorganic materials. About 1–5 per cent inorganic materials are left behind as ash when wood is burned. They consist mainly of Ca^{2+}, Mg^{2+} and K^+ cations and CO_3^{2-}, PO_4^{3-}, SiO_3^{2-} and SO_4^{2-} anions.

6. Minor components. These occur usually in cell cavities and include natural resins, oils, tannins, colouring matter, alkaloids, etc.

STRUCTURE OF WOOD

Wood, like all living organisms, is built up of individual cells which differ in size and shape according to their function: conduction of sap, mechanical support or storage of food materials. From an examination of a section of the trunk of a tree (Figs. 13 and 14) the following can be distinguished:

7. The bark. This is a protective layer of dead corky tissues which form the covering of the trunk.

8. Phloem (bast). This consists of a layer of soft moist material whose function is to conduct sugary sap from the leaves to the growing parts of the tree such as roots, cambium, etc.

9. Cambium. This consists of a thin layer of cells dividing the phloem (bark tissue) from the xylem (wood tissue).

10. Sapwood. This consists of living cells. The structure of wood formed in the early part of the growing season, known as *springwood or early wood,* is of a more open or porous nature

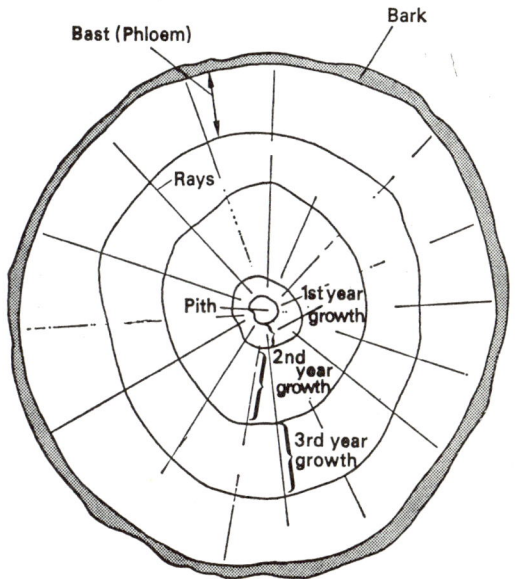

FIG. 13.—*Section through a three-year-old stem.*

than the *summerwood or late wood,* which is formed at the latter part of the season. Springwood and summerwood constitute a *growth* ring (or *annual ring* if the growth cycle takes place yearly, as in temperate climates). The function of sapwood is twofold—sap movement and food storage.

11. Heartwood. This consists of non-living cells and can often be distinguished from the paler-coloured sapwood. Deposits

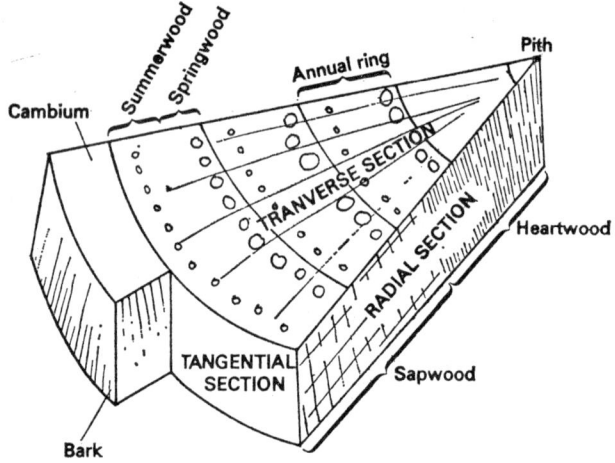

FIG. 14.—*Anatomy of timber structure.*

of colouring matter, gum, resin, tannin, etc., are commonly present in the heartwood and are responsible for its darker colour and better resistance to decay. The heartwood provides mechanical strength to the stem which supports the crown.

12. Pith (or heart). This is the soft spongy material at the centre of the section.

CLASSIFICATION OF TIMBER

Commercial timbers are conventionally classified into two main groups:

13. Softwoods (gymnosperms). Softwoods are derived from the cone-bearing trees of the *Coniferae* family, with characteristic needle-like leaves and naked seeds. They are mainly evergreens chiefly found in temperate countries. Common examples of softwoods include pine, spruce, fir, larch, cedar, hemlock, cypress, yew, sequoia.

14. Hardwoods (angiosperms). Hardwoods are derived from broad-leaved trees of the *Dicotyledon* family, with characteristic broad leaves and seeds enclosed in a seed-case. They are found all over the world. Common examples of hardwoods include oak, chestnut, ash, elm, teak (ring-porous hardwoods); beech, birch, sycamore, mahogany, iroko (diffuse-porous hardwoods).

Hardwoods are generally denser, stronger, more durable and more costly than softwoods. Some exceptions exist, for example the hardwoods willow and poplar are comparatively softer than many softwoods.

STRUCTURE OF SOFTWOOD

Visual or microscopic examination will reveal the following characteristic features of a typical softwood (Fig. 15).

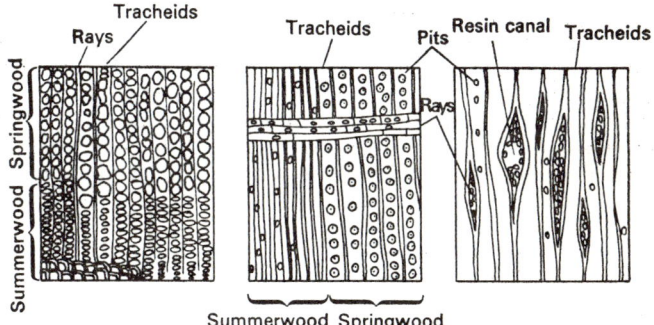

FIG. 15.—*Microstructure of a softwood.*

    (a) (*Left*) *Transverse section.*
    (b) (*Centre*) *Radial section.*
    (c) (*Right*) *Tangential section.*

15. Tracheids. These are hollow, thin, needle-shaped units usually between 2 mm and 5 mm long. They are closely packed and make up the bulk of the wood. They occur parallel to the axis of the stem or branches and provide for the conduction of sap and mechanical support for the tree. The thinner-walled tracheids in springwood with comparatively larger cavities are better suited for conduction, whilst the thicker-walled tracheids in summerwood with smaller cavities are more suited for mechanical strength.

16. Bordered pits and simple pits. These are found along the sides or walls of tracheids. Bordered pits act as valves in controlling the flow of liquids between the cells. These bordered pits are open in sapwood but are closed in heartwood—a reason why it is comparatively more difficult for liquid preservatives to penetrate through heartwood. Simple pits are merely thin windows in the wall.

17. Medullary rays (or simply "rays"). These are short, thin-walled cells which run horizontally across the grain and radiate outwards from the pith to the bark of the tree. They are used for the storage of food and are made up of storage tissue known as *parenchyma*. Pits which connect them to the tracheids are also visible.

18. Resin canals or resin ducts. These are cavities in the wood which run vertically in the stem and horizontally in the rays. They occur naturally in some species of softwood, *e.g.* larches, Douglas fir, true pines and spruces, but they are normally absent in true firs, sequoia and yew. They may be developed as a result of injury to the tree. The resin which is secreted can be beneficial in helping identification and preservation of wood.

STRUCTURE OF HARDWOOD

The structure of hardwood is more complex than that of softwood. The characteristic features of a typical hardwood are shown in Fig. 16.

19. Pores or vessels. These are tubular series of cells which run vertically along the axis of the stem and are used for the conduction of sap (compare the conducting tracheids of softwoods). Pits are present along the longitudinal walls of the vessels and are smaller than those in the walls of softwood tracheids.

20. Fibres. The woody tissue of hardwoods is made up of fibres which are narrow, vertical, spindle-shaped cells responsible for the mechanical strength of hardwoods (compare the summerwood tracheids of softwoods). It is a distinguishing feature that vessels and fibres are absent in softwoods.

21. Rays. These are cells which run horizontally across the grain and radiate outwards from the centre of the tree. In hardwoods, the rays are normally more than one cell wide, whereas in softwoods they are only one cell wide (Figs. 15(c)

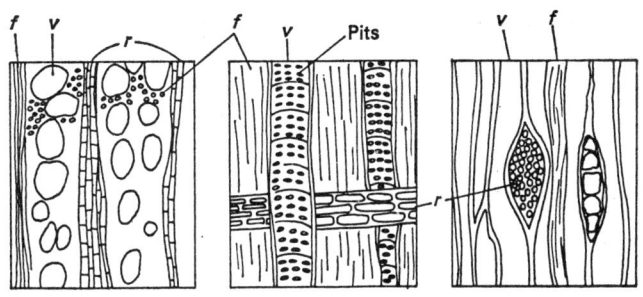

v = vessel, r = ray, f = fibres

FIG. 16.—*Microstructure of a hardwood.*

(a) (*Left*) *Transverse section.*
(b) (*Centre*) *Radial section.*
(c) (*Right*) *Tangential section.*

and 16(c)). Living cells (*parenchyma*) function as the storage tissues which occur around the pores and the rays.

22. Pits. These can be seen on the walls of pores or vessels. They are smaller and more numerous than in softwood tracheids.

RING-POROUS AND DIFFUSE-POROUS HARDWOODS

Hardwoods may be subdivided into two classes according to the arrangement of the pores.

23. Ring-porous hardwoods (e.g. oak, ash). The pores in springwood are distinctly larger than those in the summerwood and are arranged in rings or groups (Fig. 17).

24. Diffuse-porous hardwoods (e.g. beech, birch). The pores are scattered or fairly evenly distributed over the growth ring without any marked difference between those of springwood and summerwood (Fig. 18).

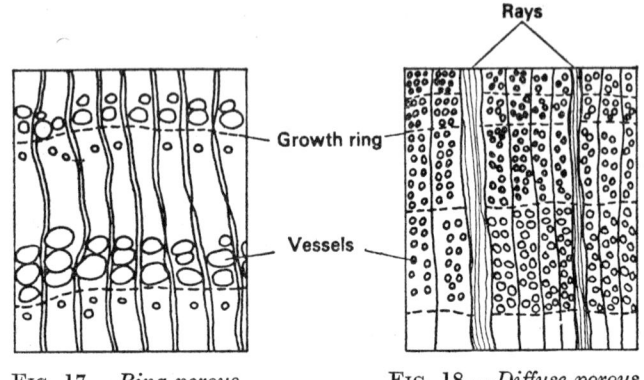

FIG. 17.—*Ring-porous hardwood.*   FIG. 18.—*Diffuse-porous hardwood.*

PROPERTIES OF TIMBER

Owing to the wide variation in species, structural features, chemical composition, water content, etc., the properties of timber may vary over a wide range. Table XXXI summarises some important properties of timber, as compared with those of metals, concrete and plastics.

SEASONING AND CONVERSION

25. Seasoning. Seasoning is the process of controlling the drying of timber. Timber from freshly-felled trees has too high a moisture content for normal use and is dimensionally unstable. For building purposes, the moisture content needs to be lowered considerably—about 24 per cent for carpentry work and about 10 per cent for joinery work. The benefits derived from seasoning of timber are improved strength, resistance to fungal decay and insect attack and dimensional stability.

There are two main methods of seasoning:

(a) *Natural seasoning* (air-drying) permits the drying of timber, which is conveniently stacked, under atmospheric

TABLE XXXI: SOME PROPERTIES OF TIMBER AND
OTHER BUILDING MATERIALS

	Timber	Metals	Concrete (gravel)	Plastics
Density (kg/m³)	320–1040	2640–11 370	2240–2480	900–2300 (cellular: 3·2–128)
Thermal conductivity $Wm^{-1} K^{-1}$	0·144	15–400 (stainless steel– copper)	1·15–1·73 (lightweight concrete: 0·5)	0·2
Thermal expansion (K^{-1})	$4·5 \times 10^{-6}$	12×10^{-6}– 31×10^{-6} (mild steel– zinc)	10×10^{-6}– 14×10^{-6} (lightweight concrete: $6·5 \times 10^{-6}$– 8×10^{-6})	7×10^{-6}– 210×10^{-6}
Strength (UTS) (Nmm^{-2})	20–110	15–617 (rolled lead– cast iron)	4	7–90 (cellular: 0·14–0·55)
Young's modulus (Nmm^{-2})	5000–9000 (softwood) 8000–18 000 (hardwood)	13 800 (lead) 96 600–13 200 (copper) 207 000 (stainless steel)	28 600	172–10 300
Fire-resistance	combustible (ignited at 221–298 °C)	non-combustible m.p. 327 °C (lead) 1453 °C (nickel)	non-combustible OPC dis-integrates at 400–500 °C	combustible (softening point 80–295 °C)
Resistance to chemicals	timber shows good resistance to alkalis and weak acids, as compared with metals and concrete			excellent (depending on type)
Moisture movement	3·4% (radial) 4·6% (tan-gential) smaller (transverse)	Nil	0·01–0·055%	Nil
Durability	liable to fungal attack	liable to corrosion, but not to fungal attack	liable to deterioration, not liable to fungal attack	excellent although may be sensitive to ultraviolet light

conditions. No artificial heat or ventilation is used, drying being mainly effected by free circulation of air through the stacked timber. This process is slow and it usually takes many months of storage before the timber is sufficiently dry for use.

(b) *Artificial seasoning* (kiln-drying) makes use of con-trolled circulation of heated air or steam through timber in special kilns. Kiln-drying is quicker but more expensive than the air-drying method. The moisture content of timber can be accurately controlled by proper and effective use of a kiln-drying process, thereby reducing the risks of possible defects such as surface hardening and distortion.

Defects due to seasoning are mainly:

(*i*) *Splits.* Lengthwise separation of timber, caused by faulty kiln operation.

(*ii*) *Warp.* A condition in which the flat, plane surface has been distorted in some manner.

An edgewise deviation is called a *crook*; a flat deviation is a *bow*; a deviation across the width is a *cup*.

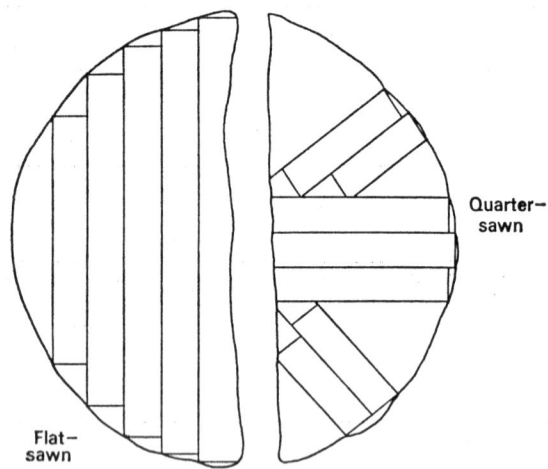

Fig. 19.—*Conversion of timber.*

26. Conversion. Conversion of timber is the process of sawing logs into convenient sizes for seasoning, marketing and use. Normally, the logs are obtained either flat-sawn or quarter-sawn, as shown in Fig. 19.

Apart from natural defects such as knots, common conversion defects arise from the cutting of timber from a round log. These are:

(i) *Shakes.* Lengthwise separations in the timber which occur between growth rings.

(ii) *Wane.* Bark or other soft material left on the edge of a board.

(iii) *Splits.* Lengthwise separation of the wood, caused by handling.

TIMBER DECAY

Timber decay can be due to various causes, such as:

(i) Mechanical wear and tear.

(ii) Adverse physical agencies, *e.g.* frost, persistent damp, prolonged heating, etc.

(iii) Aggressive chemicals, *e.g.* strong acids and alkalis, etc.

(iv) Insect attack.

(v) Fungal attack.

27. Insect attack. Attack by insects or woodworms on timber is readily recognisable by the presence of bore holes, wood dust and pellets. In the U.K. beetles are the chief pest, termites occur only in tropical countries. Beetles are not so serious as wood-rotting fungi. High humidity, warm temperature and presence of fungal decay all favour insect attack.

(a) *Life cycle of beetles* (Fig. 20). The stages in the life of a typical beetle are:

(i) *Eggs* are laid in cracks or crevices or previous exit holes.

(ii) Eggs are hatched and the *grubs (larvae)* emerge. Damage to wood is mainly due to larvae which bore their way through wood, usually leaving bore dust.

(iii) The active larva, when fully grown, changes to the dormant *pupa.*

(iv) The pupa changes into the *beetle* which emerges leaving an exit or flight hole. Within a few days, the beetle starts mating and laying eggs and the cycle is repeated again.

It will be possible to identify the type of the beetle from the examination of the beetle or larva, the bore dust and the exit hole. Some wood-destroying beetles commonly encountered in the U.K. are the house longhorn beetle, powder post beetle, common furniture beetle, deathwatch beetle and pinhole borer (*see* Table XXXII). Other timber pests include marine borers, weevils and wood wasps.

TABLE XXXII: COMMON WOOD-DESTROYING BEETLES

	House longhorn beetle (Hylotrupes bajulus)	Powder post beetle (Lyctus brunneus)	Common furniture beetle (Anobium punctatum)	Deathwatch beetle (Xestobium rufovillosum)	Pinhole borer (Ambrosia beetle)
Adult	6–76 mm long Can fly Emerging July–Sept.	Approx. 5 mm long Can fly Emerging May–Sept.	2·5–5 mm long Can fly Emerging May–Sept.	6–8·5 mm long Emerging March–June	3–6 mm long Fungus carrier
Life cycle	3–11 years or even more	Approx. 1 year or less	1–3 years or more	2–10 years	Variable, according to species
Eggs laid	Up to 200 (white and ellipsoidal)	70–220 (white and slender)	Up to about 80 (white, lemon-shaped)	40–70 (white, lemon-shaped)	Laid in tunnels cut by females
Damage due to	Larva or grub	Larva or grub	Larva or grub	Larva or grub	Adult beetle

Timber attacked	Seasoned sound timber; usually softwood Sapwood only	Seasoned and partly seasoned sound hardwood Mainly sapwood	Seasoned hardwood and softwood—sound or decayed Mainly sapwood	Old decayed hardwood; softwood rarely Sapwood and heartwood (*dry* timber not attacked)	Unseasoned sound hardwood and softwood Heartwood and sapwood
Exit holes	Oval, 6–9·5 mm diameter	Circular, approx. 1–1·5 mm diameter	Circular, approx. 1·5 mm diameter	Circular, approx. 3 mm diameter	Circular, approx. 0·5–3 mm diameter
Bore dust	Compact cylindrical pellets and powder	Fine talcum-like powder	Egg-shaped pellets	Coarse bun-shaped pellets	Absent, but tunnels darkly stained
Preventive measures	Early removal of timber after felling in winter	Reject all infected timber before shipping, damage continues after manufacture Sterilise before use Periodic inspection	Use of well-seasoned timber free from sapwood	Specialists' job required Removal of infected wood and bore dust Application of insecticide	Early removal of timber after felling Attack ceases when timber is seasoned

(b) *Eradication of wood-destroying beetles.* As soon as exit holes are found it is necessary to check whether the wood-worm attack is still active, for example, by prying into these holes with a pointed object and looking for signs of bore dust. If the infestation is widespread, specialists should be called for free survey and assessment of the extent of attack.

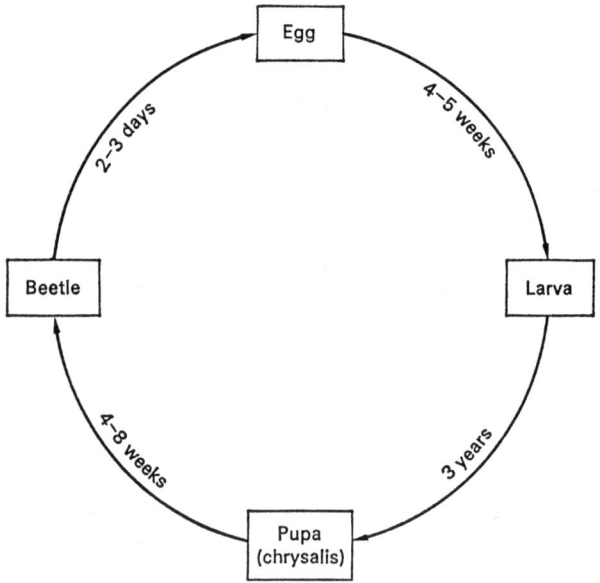

Fig. 20.—*Life cycle of common furniture beetle.*

Severely attacked timber (including that showing bore dust) should be carefully removed and replaced with pretreated timber. Timbers not replaced should be generously sprayed on all sides with an appropriate timber preservative. Periodic inspection (*e.g.* annually) to the treated timber and the adjoining timber is necessary until no more exit holes or bore dust appear. Specialist firms normally give a twenty-year guarantee for timbers replaced should any recurrence of woodworm occur.

28. Fungal attack. Attack by fungi, for example by dry rot fungi, can be very serious, and treatment has to be more drastic than in the case of woodworm attack. Not all specialist firms are prepared to give a twenty-year guarantee after treatment.

(*a*) *Life cycle of fungi* (Fig. 21). Fungi belong to a class of vegetable kingdom which differs from "green" plants in that they are unable to produce their food material by the process of photosynthesis. They therefore derive their nourishment from organic materials (such as timber) and exist as *parasites*

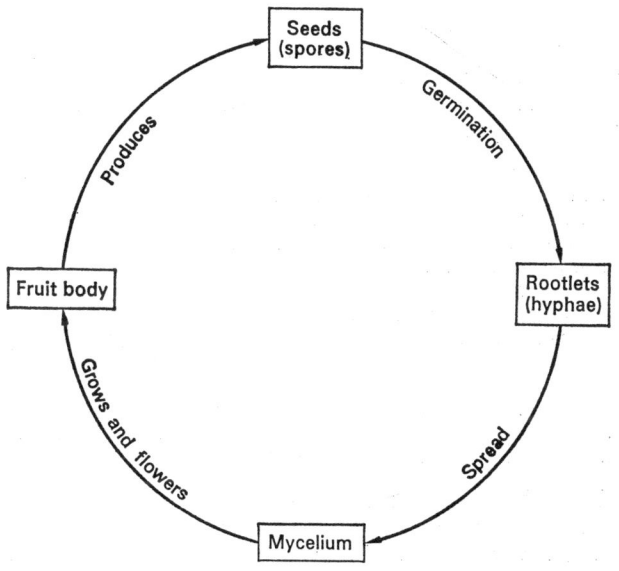

Fig. 21.—*Life cycle of a fungus.*

on living plants or animals or as *saprophytes* on the dead remains of plants or animals. The stages in the life of typical fungus are:

(*i*) *Seeds or spores* of the fungus fall on a piece of wood.

(*ii*) With the right conditions of humidity and temperature, germination of these spores takes place, sending fine *rootlets or hyphae* into the wood.

(*iii*) These hyphae grow and spread into fluffy masses of *mycelium*.

(*iv*) When the fungus is fully grown, it flowers and gives rise to *fruit bodies,* which in turn produce more spores.

(*b*) *Conditions for fungal growth.*

- (*i*) Source of infection (*i.e.* presence of spores).
- (*ii*) Presence of food supply (*i.e.* timber, coal or other organic materials).
- (*iii*) Suitable temperature (*i.e.* in the range 4–38 °C).
- (*iv*) Suitable moisture content (*i.e.* 20–40 per cent).
- (*v*) Availability of oxygen.

Therefore, fungal growth is unlikely in situations where very high or very low temperature exists or where low moisture content or a saturated condition (ponding) exists.

(*c*) *Types of fungi.* There are basically two main types of wood-rotting fungi:

(*i*) *Dry rot fungus (e.g. Merulius lacrymans).* This is the true dry rot and is most destructive because this fungus has the ability to produce hyphae which can grow over brickwork, concrete and metal and can even penetrate plaster and soft brickwork and mortar, in search of fresh timber which it will then attack.

(*ii*) *Wet rot fungus. Coniophora cerebella* (cellar fungus) requires more moisture (25 per cent or more) than dry rot fungus and is not so serious as the infection stops as soon as dampness is cured.

(*iii*) *Poria vaillantii* (pore fungus) occurs in rather wet conditions, such as in coalmines. Its growth can be checked by drying out the infected areas of timber.

Table XXXIII summarises some of the main characteristics of common wood-rotting fungi.

(*d*) *Detection and eradication of fungal decay.* Early detection of fungal decay is seldom possible. The attack usually starts in a most inconspicuous place where moisture can gain access, *e.g.* through leakages in roof (defects in valley gutters, down drains or gulleys), from defects in plumbing systems (leaky taps, central heating radiators), from lack of, or defects in, damp-proof courses, or as a result of persistent condensation (kitchen, bathroom). Often the fungal attack is only apparent when it is at an advanced stage, when more drastic and costly treatment is required. The symptoms of fungal decay in timber are shown by the change in colour (darker colour), the change in strength and density (brittleness and lightness) and the change in odour (usually a musty smell). If the attack is widespread mycelium and strands of hyphae can be observed. It is sometimes possible for someone allergic to minute spores of fungus to detect an outbreak of dry rot before it becomes widespread.

Once detected, the type of decay must be ascertained so

TABLE XXXIII: TYPES OF WOOD-ROTTING FUNGI

| | DRY ROT FUNGUS (*Merulius lacrymans*) | WET ROT FUNGUS | |
		Coniophora cerebella (*cellar fungus*)	*Poria vaillantii* (*pore fungus*)
Spores	Yellowish, elliptical 8–10 × 5–6 μm in size	Yellowish brown, ovoid 11–13 × 7–8 μm in size	White, elliptical 6 × 3·5 μm in size
Occurrence	In damp, unventilated places where moisture content of timber exceeds 20% Can spread to dry timber	In very damp conditions (*e.g.* bathroom, cellars or places associated with leaks and condensation) Does not spread to dry areas	In very wet conditions (*e.g.* coalmines) Only limited power of spreading beyond damp area
Wood attacked	Softwoods and occasionally hardwoods	Softwoods and hardwoods	Softwoods
External appearance	Mycelium-white, fluffy with yellow patches, or in drier conditions matted grey skin with tinges of purple	Generally no external growth, sometimes blackish brown strands on surface	Hyphae and mycelium —white and soft
Fruit body	Pancake-like, flattened, tough but fleshy with edges and brick-red centre	Rare in buildings Thin flat plate, olive green or brown in colour	White or creamy white, plate-shaped, covered with fine pores
Effect on wood	Dry and brittle, generally light brown in colour Deep cracks along and across grain	Dark brown cracks along grain (early stages of attack) and across grain (more advanced stage of attack), but cubes are less pronounced than with *Merulius*	Light brown cracks along and across grain Cubical cracks are less pronounced than with *Merulius*

that appropriate treatment can be prescribed. Whatever the cause, the source of dampness is to be traced and cured. In the case of wet rot, treatment is comparatively simple—the infected timber is treated copiously with timber preservative or, if in an advanced stage of attack, the timber can be replaced by a pretreated timber. However, if dry rot is suspected or confirmed, more drastic steps must be taken:

(*i*) Full extent of the outbreak to be determined and exposed.

(*ii*) Source of dampness to be traced and cured.

(*iii*) All mycelial strands to be removed and burnt.

(*iv*) All decayed timber to be removed, together with about 0·5 m of apparently sound timber beyond the decayed regions.

(*v*) All wall surfaces (masonry or brickwork) to be scraped clean and irrigated with preservatives (such as sodium orthophenylphenate or sodium pentachlorophenate).

(*vi*) Adjacent timber or brickwork to be treated with preservatives for an area of approximately 1 m beyond the affected spot.

(*vii*) New sterilised and preserved timber to be put in place.

(*viii*) Frequent inspection to be carried out to monitor success of treatment.

There is no certainty that the eradication is complete or the treatment is complete and effective, as the extent of infection is difficult to assess. There is always a possibility of a recurrence of attack or infestation. It may seem costly to call in a specialist firm to do the job, but it certainly will save time, money and worry in the long run.

It is worth mentioning that apart from the use of preservatives which are very effective, treatment by heat (blow-lamp and radio-frequency methods) is also effective. Blow-lamp heating involves some fire risk and the radio frequency methods, so far, have proved to be too difficult and dangerous to be of any commercial value.

TIMBER PRESERVATION

Sapwood of all species of timber in damp environments is always prone to decay and insect attack. Heartwoods of some timbers (*e.g.* afromosia, greenheart, iroko or teak) contain resins which are self-preservative and render them very resistant against attack by fungus or insects. Other timbers (*e.g.* ash, beech, whitewood) are not so durable, especially in damp surroundings. Much of the timber used nowadays contains a proportion of sapwood and therefore preservative treatment is necessary, particularly in situations where dampness is likely to occur. Timber preservation depends on the effectiveness of the preservative and on the efficiency of the method of treatment.

29. Timber preservatives. The main considerations affecting the choice of preservatives are:

(a) *Toxicity.* It should have maximum toxicity to fungi and pests, but minimum toxicity to human beings and animals.

(b) *Permanence.* Ideally it should be effective against decay for an indefinite period of time and must therefore be resistant to leaching, exudation, evaporation and volatilization of the active ingredients.

(c) *Availability and cost.* It should be readily available and relatively inexpensive.

(d) *Chemical stability.* It should not corrode metals in contact with treated timber and should not increase fire risk.

(e) *Decorative effect.* It should not adversely affect the subsequent application of paint or other decorative finishes.

No known preservative, however, has yet been produced which can satisfy all these requirements.

According to BS 1282, timber preservatives may be classified in three main types:

(i) Type TO—Tar oil types.

TO 1: Coal tar creosote BS 144 for pressure impregnation.
TO 2: Coal tar oil types BS 3051 for brush application.

Creosote is derived from fractional distillation of coal tar. It is an effective and inexpensive timber preservative, resistant to leaching and generally not corrosive to metals. On the other hand, it has certain disadvantages—a characteristic odour, discoloration of adjacent materials and plasterwork, difficulty of painting over treated areas. It is unsuitable for joinery work but suitable for such work as fencing, marine piling poles and wooden bridges.

(ii) Type OS—Organic solvent types.

OS 1: Chlorinated naphthalenes and other chlorinated hydrocarbons.
OS 2: (a) Copper naphthenate.
(b) Zinc naphthenate.

OS 3: Pentachlorophenol and its derivates.
Mixtures of OS types.

These consist of solutions of preservative substances in organic solvents (such as petroleum oils). They are resistant to leaching (hence suitable for exterior and interior use), non-corrosive to metals, non-staining, but are expensive, inflammable and have a characteristic odour. Treated wood will not swell or distort and can be painted over satisfactorily.

Important examples are:

Cuprinol (copper and zinc naphthenate).
Copper pentachlorophenate (copper PCP).
Benzene hexachloride (BHC).
Tributyl tin oxide.
Dieldrin (chlorinated hydrocarbon)—at present banned by the government.

(*iii*) Type WB—Water-borne types.

WB 1: Copper/chrome.
WB 2: Copper/chrome/arsenic.
WB 3: Fluor/chrome/arsenate/dinitrophenol.
WB 4: Others, *e.g.* copper sulphate, mercuric chloride, sodium fluoride, sodium pentachlorophenate, zinc chloride and organic mercurial derivatives.

These consist of preservative substances dissolved in water. They are odourless, inexpensive, non-inflammable and can be painted over when dry. With the exception of WB 4 they are virtually non-corrosive to metals and are resistant to leaching.

30. Methods of preservation. The principle of preservation involves the introduction of preservatives to an appreciable depth in the timber. Some species (*e.g.* beech and ramin and the sapwood of most species) can be readily impregnated, while others (*e.g.* Douglas fir and oak heartwood) are not easy to impregnate even with prolonged high-pressure methods.

The efficiency of the method depends on the porosity of the timber and on the depth of penetration by the preservatives.

Prior to treatment, the timber must be properly dried (the moisture content below 22 per cent normally) and made free from dirt and grease. For some timbers (*e.g.* Douglas fir and spruce) which are very resistant to penetration by preservatives, incisions can be made at regular intervals in the surface of the timber by a special machine.

The main methods of preservation include:

(a) *Brushing and spraying*. The easiest but the least effective method, which can be used with all types of preservatives. Suitable for treatment of timber *in situ*. The penetration obtained seldom exceeds 1·5 mm. Preservatives should be applied as liberally as possible, creosote and other tar oil types should be preferably applied hot at a temperature of about 38 °C.

(b) *Dipping*. This method involves submerging timber in a preservative, from a few seconds (not less than 10) to a few minutes. The longer the period of immersion the better the penetration. Oily-type preservatives should be preferably heated. The timber to be treated should be clean, reasonably dry (moisture content below 25 per cent) and free from frost, snow and water from recent rain. This method is slightly more effective than brushing.

(c) *Steeping*. The period of immersion in a cold preservative (preferably water-soluble types) varies from a few hours to a few days. Deeper penetration is obtained and hence this method is more effective than brushing, spraying and dipping. Posts and timber for similar uses should be steeped at least 0·3 m above ground level.

(d) *Hot and cold open tank process*. The timber is submerged in a tank of preservative, which is then heated to a suitable temperature (*e.g.* 80–90 °C for oily-type preservatives and up to 66 °C for aqueous solutions of certain metallic salts). This method is as effective as the pressure processes in the treatment of sapwood and permeable heartwood. Suitable for treatment of gateposts, fencing, etc.

(e) *Pressure processes* (BSS 913, 407). Unlike all the previous methods, which are non-pressure processes, here the preservatives are impregnated by application of pressure; this gives deepest penetration and is the most effective method of treatment for most timbers. Special plant is required.

There are two principal types of pressure treatment:

(i) *Full-cell process*. The timber is sealed in a pressure cylinder or autoclave. An initial vacuum is applied (in order to remove air and surface moisture from timber) and held for a period of 30 minutes to 1 hour. The preservative (usually WB) is then introduced to fill the cylinder. The vacuum is released and hydraulic pressure is applied and maintained for a few hours, the length of time depending on the type of timber and preservative. The pressure is released and the preservative withdrawn, followed by a short final vacuum to dry the surface for convenience of handling.

(*ii*) *Empty-cell process*. This differs from the full-cell process in that no initial vacuum is used. As a result air remains in the cells and after treatment the cell cavities of timber are only partially filled with preservatives. This method is normally confined to treatments with creosote or tar-oil types.

(*f*) *Diffusion process*. Green (unseasoned) timber is dipped in a hot solution of water-borne boron preservative and then close-piled for several weeks, when diffusion of the boron preservative into the timber occurs before drying takes place. The process is unsuitable for timber which will be exposed externally, due to leachability of the water-soluble salts.

TIMBER PRODUCTS

Timber, because of its fibrous nature, is highly anisotropic—both tensile and compressive strengths are much greater along the grain than across the grain, whilst the shear strength is lower along the grain than across it. For this reason, the use of timber as tension members of a structure is restricted.

31. Plywood. The discovery of plywood is one way of overcoming the problem of anisotropy.

Plywood is a composite form of wood made up of three or more thin layers of wood (called plies or veneers) which are bonded together and arranged such that the crossing of the grain at right angles in adjacent veneers tends to balance the strength in both directions. In this way, greater strength and lightness together with high dimensional stability can be obtained. Plywood compares favourably with most materials in construction work (Table XXXIV).

Plywood can be obtained in large-sized sheets and in varying thickness (3–25 mm).

Timber suitable for manufacture into plywood includes alder, beech, birch, Douglas fir, mahogany, oak, walnut and certain West African timbers such as makore, sapele, seraya and utile. The types of adhesive used for bonding the veneers or plies include animal and casein glues (for ordinary purposes), synthetic resin glues (for special purposes), *e.g.* phenol-formaldehyde, urea-formaldehyde, resorcinol-formaldehyde, melamine-formaldehyde.

TABLE XXXIV: STRENGTH OF PLYWOOD AND OTHER MATERIALS

Materials	Strength/weight ratio
Birch plywood	8·5
Douglas fir plywood	6·0
Aluminium alloy	1·5–8·4
Mild steel	3·4

Relevant BS specifications are:

> BS 565: 1972 Glossary of terms relating to timber and woodwork.
> BS 1203: 1963 Synthetic resin adhesives (phenolic and aminoplastic) for plywood.
> BS 1455: 1972 Plywood manufactured from tropical hardwoods.
> BS 3493: 1962 Information about plywood.
> BS 1088 and
> 4079: 1966 Plywood for marine craft.

32. Blockboard and laminboard. *Blockboard* is made up of a solid core consisting of blocks up to 25·4 mm wide sandwiched between two veneers, arranged so that the grain in the veneer is at right angles to that of the core blocks. It is available in various thicknesses (12–43 mm and 38–48 mm) and in sizes up to 1·65 × 3·66 m.

Laminboard is similar, except that the core strips are narrow (up to 7·1 mm) and are continuously glued.

Relevant BS specifications are:

> BS 3444: 1972 Blockboard and laminboard.
> BS 3583: 1963 Information about blockboard and laminboard.

Mainly used in furniture and fittings.

33. Particle boards (wood chipboards). These are made from particles of wood and/or other lignocellulosic material bonded with synthetic resin and/or other organic binder.

Available in various sizes and thicknesses, the boards can be veneered with wood or plastic veneers. Main uses include floorings, wall and ceiling linings, concrete formwork linings, roof decking, furniture.

Relevant BS specifications are:

> BS 1811: 1969 Methods of test for wood chipboards and other particle boards.
> BS 2604: 1970 Resin-bonded wood chipboards.

34. Fibre-building boards (BS 1142). These are made from wood or other vegetable fibres, with varying range of properties. Three main types are listed in BS 1142:

(a) *Insulating fibreboards.* These have generally low densities and low thermal conductivities. They are available in various forms such as:

(i) Standard insulating boards.
(ii) Bitumen-bonded insulating boards.
(iii) Bitumen-impregnated boards.
(iv) Acoustic boards.
(v) Flame-retardant boards.

(b) *Wallboards.* Slightly denser than insulating boards (not exceeding $480 \cdot 6$ kg/m^3) and used mainly as room linings and underlays for sheet floorings. They are available in various forms—homogeneous fibre wallboard or building board.

(c) *Hardboards.* Denser than wallboards and insulating boards:

(i) Medium hardboards: density $480 \cdot 6$–$800 \cdot 9$ kg/m^3.
(ii) Standard hardboards: minimum density 881 kg/m^3.
(iii) Tempered hardboards: minimum density $961 \cdot 1$ kg/m^3.

Main uses include wall and ceiling linings, underlays for floors, linings for concrete formwork and in special forms for sound absorption purposes.

PROGRESS TEST 7

1. In the case of a tree, what are the functions of the root, the trunk and the crown? (1)

2. What is the chemical composition of wood? (2–6)

3. Describe, with the aid of a labelled diagram, the macro-structure of wood. (7–12)

4. What are the basic differences between a typical softwood and a typical hardwood? Name five examples of softwoods and hardwoods respectively. (13–14)

5. What are the main microstructural features which distinguish between a softwood and a hardwood? (15–24)

6. Compare the characteristic properties of timber with other traditional building materials. (Table XXXI)

7. Distinguish between seasoning and conversion of timber. What are the main defects which may occur as a result of seasoning and conversion of timber? (25–26)

8. State the main factors which may cause timber decay. (27–28)

9. Outline the life cycle of (i) a wood-destroying beetle, (ii) a wood-rotting fungus. (27(a), 28(a))

10. What conditions favour insect and fungal attack on timber? (Table XXXII)

11. Describe the appearance of timber attacked by insect and by dry rot respectively. (Tables XXXII, XXXIII)

12. Give a brief account of the procedures to be taken towards eradication of dry rot. (28(d))

13. What considerations are necessary for the choice of a timber preservative? (29)

14. Name the three main classes of timber preservatives. List their characteristic properties in each case. (29)

15. Name two species of timber which lend themselves to easy impregnation by preservatives and two species which do not. (30)

16. Compare the principles and efficiencies of the various methods of timber preservation. (30)

17. What is meant by the term "anisotropy" in the case of timber? (p. 126)

18. What is plywood? How can plywood overcome the problem of anisotropy in timber? (31)

19. Name the various timber products available commercially, and mention their main uses. (31–34)

EXAMINATION QUESTIONS

1. Describe the structure and growth of a tree trunk. Explain the structural differences between softwoods and hardwoods.

Outline the difference in behaviour between thin sections and large sections of wood, when exposed to fire, and describe two treatments which can be used to give a Class 1 rating for surface spread of flame.

(P.S.B. B.Sc. S.E.)

2. (a) With reference to timber technology, explain the difference between:

(i) hardwood and softwood,

(ii) conversion and seasoning,

(iii) heartwood and sapwood.

(b) Describe the manufacture and properties of plywood.

(S.O.E. Grad. Exam. C.E.)

3. Describe the ways of converting logs into timber sections, and explain how the different types of cut affect the properties in use of the timber sections.

Discuss:

(a) the types of pore distribution in hardwoods, and their effect on the properties of the timber.

(b) the nature and function of tracheids and parenchyma.

(P.S.B. B.Sc. S.E.)

4. (a) Describe the structure of a typical softwood, including both growth defects and seasoning defects.

(b) Describe a named insect attack and a named fungal attack on softwood; indicate possible methods of control of the infection.

(P.S.B. H.N.D. Bldg.)

5. What are the essential properties of a timber preservative?

Give an account of *three* different methods of timber preservation, indicating the type of preservatives used in each method.

(S.O.E. Grad. Exam. C.E.)

6. (a) Give a critical account of the properties and uses of softwood as a building material, as compared with concrete and plastics.

(b) Briefly discuss the following methods of preservation of constructional timber:

(i) Impregnation process,

(ii) Open-tank process.

(P.S.B. B.Sc. S.E.)

7. "The formation of a practically useful adhesive is more an art than a science; nevertheless, there are certain guiding principles which can be successfully invoked."

(a) Name, describe and explain TWO such scientific principles and demonstrate their implementation in a commentary on adhesives which will include, with examples, their classification, application and setting actions.

(b) Briefly outline the production of plywood and chipboard.

(P.S.B. B.Sc. Bldg.)

8. (*a*) Discuss the relative merits of timber, steel and reinforced concrete as materials for the structural frames of large buildings.

(*b*) Show by means of detailed sketches a suitable method for connecting a ring beam to a column in each of the following materials:

    (*i*) precast reinforced concrete,
    (*ii*) universal section steelwork,
    (*iii*) timber.

(I.O.B. Assoc. Part 1)

FURTHER READING

Desch, H. E., *Timber, its structure and properties*, Macmillan, **1973.**
Findley, W. P. K., *Timber Pests and Diseases*, Pergamon, **1967.**
Scott, G. A., *Deterioration and Preservation of Timber in Buildings*, Longmans, 1968.

CHAPTER VIII

BITUMINOUS MATERIALS

As early as 3800 B.C. the use of bitumen as a bonding and waterproofing material was known. In earlier times the bitumen used came from natural deposits. Nowadays it is obtained mostly from artificial sources such as petroleum refinery industries.

BITUMEN AND ASPHALT

1. Nomenclature and definition of terms. BS 892: 1967—Glossary of highway engineering terms—differentiates between the two as follows:

> *Bitumen:* A viscous liquid, or a solid, consisting essentially of hydrocarbons and their derivatives, which is soluble in carbon disulphide; it is substantially non-volatile and softens gradually when heated. It is black or brown in colour and possesses waterproofing and adhesive properties. It is obtained by refinery processes from petroleum, and is also found as a natural deposit or as component of naturally occurring asphalt, in which it is associated with mineral matter.

> *Asphalt:* A general term for certain mixtures of asphaltic cement and mineral matter.

NOTE: The American term "Asphalt" is equivalent to the term "bitumen" used in the U.K.

2. Properties of bitumen and asphalt.

(*a*) *Bitumen:* A black substance which is opaque, matt or glossy in appearance. It is described as having a "dead" smell on heating as opposed to the aromatic smell of hot tar. It is insoluble in water but soluble in carbon disulphide.

(Tar is insoluble in carbon disulphide.) It has good bonding and waterproofing properties. It softens with the rise in temperature and its liquid viscosity falls as the temperature is raised. Chemically it is very complex with a molecular weight range of 900–5000. It is considered to be mixtures of hydrocarbons which are predominantly aliphatic in nature.

(b) *Asphalt:* This is a composite material consisting of mineral aggregates bonded together by bitumen. It is usually plastic in character, strongly adhesive and waterproof. Its plasticity is affected by heat and overheating may cause oxidation and loss of volatile components resulting in hardness and brittleness and some loss in adhesive properties. At ordinary temperatures it is very resistant to water, alkalis, brine or sulphates. It is, however, attacked by weak acids, hot alkalis, mineral oils, grease, vegetable fats, hot sugar solutions, milk and dairy products.

3. Sources of bitumen.

(a) *Natural sources.*

(i) *Lake asphalt.* The main source is Trinidad Lake Asphalt (Trinidad in the West Indies), apparently discovered by Christopher Columbus (1498) and Sir Walter Raleigh (1595). Asphalt quarried from this source is a hard black substance. After refinement at the lakeside it is exported under the name of Epuré. It is much too hard for direct use and therefore it is usually softened or fluxed with a semi-solid or fluid petroleum residue before incorporation into an asphalt mixture. Its composition is fairly constant—a typical analysis shows the following result:

Bitumen soluble in carbon disulphide	53–55%
Mineral matter	36–37%
Organic insoluble	9–10%

(ii) *Rock asphalt.* Natural rock asphalt deposits are found all over the world—in France, Germany, Italy, Sicily, Switzerland, and in Kentucky, Texas, California (United States), Altrabaska (Canada), etc. The rock is a fine-grained limestone impregnated with bitumen up to 11 per cent. As imported it must have a minimum bitumen content of 6 per cent according to the appropriate BS. It is normally exported in lump form as quarried, being ground in this country and enriched by the addition of bitumen.

A typical analysis of such a rock from St. Jean de Maruéjols in France is as follows:

Bitumen	10·5%
Calcium carbonate	84·2%
Magnesium carbonate	2·8%
Iron and aluminium oxides	1·6%
Silica	0·4%
Sulphur trioxide	0·5%
	100·0

(b) *Artificial sources*—distillation of crude petroleum. Distillation of crude petroleum (crude oil) produces a very wide range of products. Light fractions (such as gasoline, kerosine and gas oil) are first obtained below 350 °C under atmospheric pressure. Heavier fractions (such as lubricating oils and bitumen) are not heated above 400 °C for risk of "cracking" (decomposition) but are refined by the use of reduced pressures and steam injection in the fractionating column. Bitumen settles to the bottom of the column where it can be withdrawn.

A typical analysis is as follows:

Bitumen soluble in carbon disulphide	99·5%
Mineral matter	trace
Organic insoluble	nil

4. Modified forms of bitumen. Three methods are available to reduce the viscosity of bitumen prior to application:

(a) Heating.

(b) Using a suitable solvent for bitumen (*cut-back bitumen*).

(c) Using an emulsifying agent (*bituminous emulsions*).

Solvents used are mostly organic liquids such as carbon disulphide, benzene, petroleum oils, naphtha, etc.

Emulsifying agents normally used are soaps, alkaline casein solutions, bentonites, etc.

(a) *Cut-back bitumen.* This is defined as a bitumen whose viscosity has been reduced by the addition of a suitable volatile dilutent.

Its manufacture involves dissolving bitumen in a suitable solvent (*e.g.* petroleum oil) by using either the *batch process* or the *continuous process*. The *batch process* requires a rigid system of manufacture, supervision, testing and storage—making it rather uneconomical.

Cut-back bitumen is widely used in:

(*i*) Road building and maintenance.
(*ii*) Preparation of sealing, sound insulating and water-proofing materials.
(*iii*) Preparation of paints.
(*iv*) Preparation of plastics and mastics.

(*b*) *Bitumen emulsions*. An emulsion is a relatively stable suspension of a liquid which is minutely subdivided (the *dispersed phase*) and evenly dispersed in another liquid in which it is not soluble (*the continuous phase*).

A bitumen emulsion is obtained by emulsifying the bitumen (dispersed phase) in an aqueous solution (continuous phase). Dispersion can be facilitated by use of a simple paddle-type stirrer, or, better still, by use of a rugged type of colloid mill.

Bitumen emulsions must be perfectly homogeneous and able to withstand storage and shipping. They must not be subjected to temperatures below the freezing point of water, otherwise coagulation of bitumen particles occurs owing to the freezing of the aqueous solution.

Typical road emulsions contain 50–60 per cent bitumen, usually of penetration grade 180/220 or 280/320.

Their main uses include:

(*i*) Road building and maintenance.
(*ii*) Soil stabilisation (to prevent soil or clay from taking up sufficient water to lose stability).
(*iii*) Emulsified bitumen–earth mixtures in construction of buildings.
(*iv*) Waterproofing masonry and concrete foundations.
(*v*) Waterproofing and insulating roofs.
(*vi*) Protection to wood and metal surfaces from atmospheric deterioration and corrosion.

In a bitumen emulsion the minute particles of bitumen normally carry electrostatic charges. Hence there are two types of emulsion according to the nature of the charge carried:

Anionic emulsions (BS 434: 1973, BS 2542: 1960) in which the bitumen droplets carry *negative* charges and the aqueous solution is alkaline (*pH* 9–11). They are unstable on addition of acid or acidic salts. Emulsifying agents are usually soaps and surface active materials.

Cationic emulsions (BS 434: 1973) in which the bitumen droplets carry *positive* charges and the aqueous phase is acidic. Cationic emulsion must not be used with anionic emulsions for the obvious reason that immediate coalescence will occur. Most aggregates (with the exception of limestone) when wet carry a negative charge at the surface and therefore have strong attraction with cationic emulsions.

Emulsifying agents are usually certain fatty diamine salts and quaternary ammonium salts.

TAR AND TAR PRODUCTS

Tar is defined as a viscous liquid produced by the destructive distillation or carbonisation of coal, wood, shale, etc. It possesses adhesive properties and is obtained as the condensible distillate, whereas *pitch* is the residue obtained from the distillation of tar.

Pitch, which is liquid when hot and solid when cold, can be classified into three types: hard (softening point 85 °C), medium soft (softening point 73–82 °C) and soft (softening point 65 °C).

Road tar (BS 76: Tars for road purposes). This is coal tar which has been so treated as to conform to a specification defining its properties for use as a binder in road construction.

Unlike petroleum bitumen, tars are mixtures of hydrocarbons which are predominantly aromatic and have molecular weights ranging from 150 to 3000.

There are four main types of coal tar produced, depending on the method of production:

(a) Coke oven (U.S.A.—(approx.) output 8 gallons tar per ton of coal).
(b) Vertical retort (approx. output 14–17 gallons tar per ton of coal).
(c) Horizontal retort (U.K.—approx. output 9–11 gallons tar per ton of coal).
(d) Low-temperature tar (unsuitable for refining to road tar).

Three stages are involved in the production of road tars:

(*i*) Carbonisation or destructive distillation of coal
→ crude coal tar.
(*ii*) Refining or distillation of crude coal tar
→ oil fractions and pitch.
(*iii*) Blending of distillation with distillate oil fractions
→ desired road tars (type A and type B) (*see* Fig. 22).

Road tar (type A) is mainly used as surface dressings and base

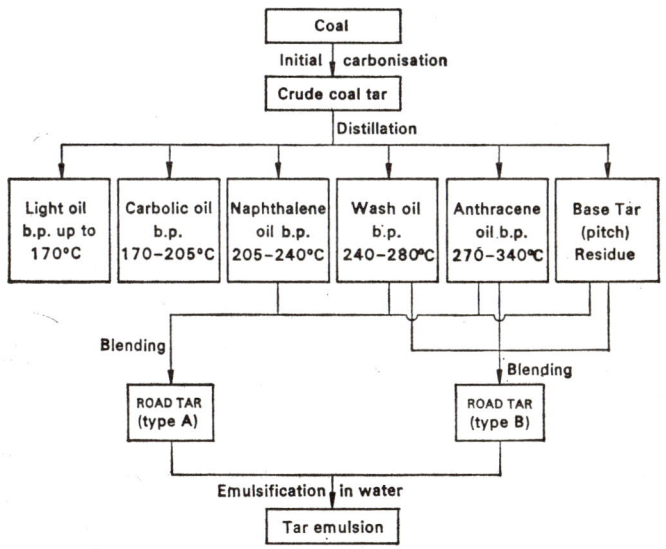

FIG. 22.—*Production of road tar.*

courses. Road tar (type B) is mainly used in tar macadam and as wearing courses and carpets.

By emulsifying road tar in water, tar emulsion is obtained. Tar emulsion is, however, seldom used for road purposes in the U.K.

TESTS FOR TARS AND BITUMENS

5. Measurement of flow properties.

(*a*) *Penetration test* (for bitumen). By measuring the penetration (in units of 0·1 mm) of a standard needle through a sample under a load of 100 g applied for 5 sec. at 25 °C (*see* Fig. 23). See BS 4691, BS 4698 for details.

(*b*) *Viscosity tests.*

(*i*) *Standard tar viscometer* (*STV*) (for tar and cut-back bitumen). Based on the principle of flow of liquid through a

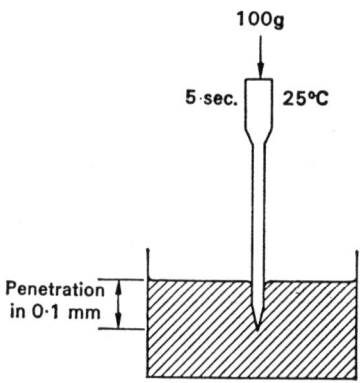

Fɪɢ. 23.—*Penetration test for bitumen.*

tube. By measuring the time taken for a specified amount of sample (50 ml) to flow through the orifice of a standard cup (10 or 4 mm) at the test temperature (Fig. 24).

(*ii*) *Equi-viscous temperature viscometer* (*ETV*) (especially for tar). A type of rotational viscometer (Fig. 25). By measuring the temperature (called *EVT*) at which a sample of tar must be heated to produce an efflux time of 50 sec. The *EVT* can be alternatively deduced from the viscosities obtained by *STV* method. Relevant information from BSS 4459, 4693, 4708.

(*c*) *Softening point test* (ring and ball test). By measuring the temperature at which a disc of bitumen softens sufficiently to allow a steel ball (9·5 mm ($\frac{3}{8}$ in.) diameter and weighing 3·50 g) initially on the surface to fall a specified

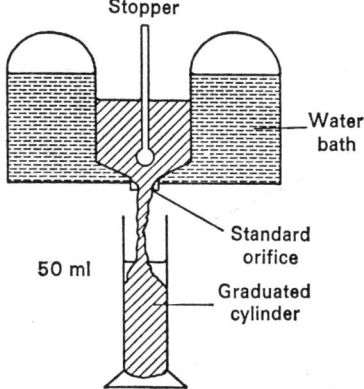

FIG. 24.—*Standard tar viscometer.*

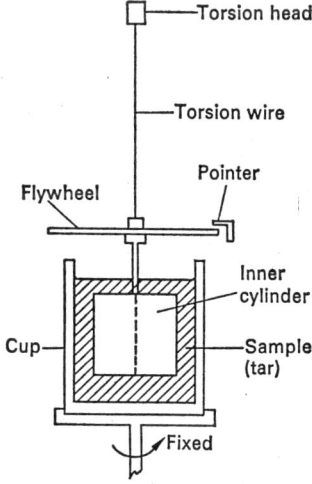

FIG. 25.—*Equi-viscous temperature viscometer.*

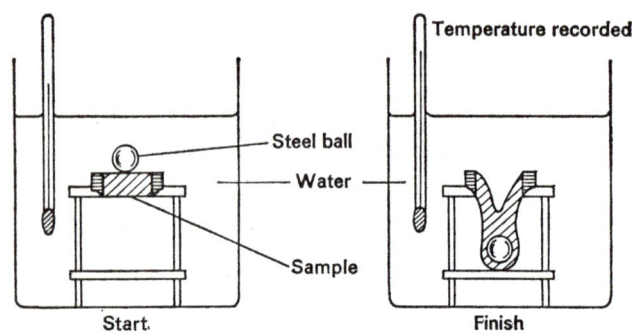

FIG. 26.—*Softening point test (ring and ball test).*

distance (25 mm) through the disc (Fig. 26). Relevant information from BSS 4692, 5094.

(*d*) *Ductility test.* Ductility is determined by measuring the distance that a briquette of bitumen necked to a cross-section of 100 mm² will stretch without breaking when elongated at a rate of 50 mm/min. at 25 °C (Fig. 27). See, however, BS 4710.

6. Specific gravity test. Similar to "pyknometer" or specific gravity bottle method. Details from BS 4699.

Useful in determining the percentage of voids in mechanically-designed mixtures of bitumens and mineral aggregates (bituminous surfacings). Bitumen has a specific gravity of 1;

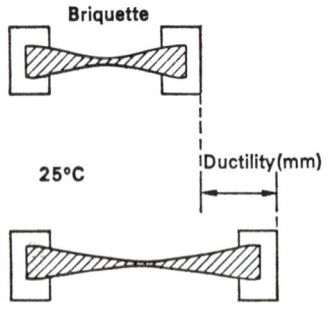

FIG. 27.—*Ductility test.*

tar of 1·10–1·15 (vertical retort), 1·18–1·25 (horizontal retort). Normal temperature 15·5 °C, but other temperatures can be used.

7. Flash and fire point test. A sample of the binder is heated at a uniform rate (5–7 °C/min.) and a small flame (4 mm (0·16 in.) diameter) is applied periodically across the surface of the heated sample until a flash first appears (this is taken as the *flash point*). If the heating is continued until the sample burns for at least 5 sec., the temperature is called the *fire point*. Relevant information from BSS 4688, 4689, 4703.

8. Composition tests.

  (a) *Distillation tests* (BSS 4349, 4453, 4700).

    (i) For tar: air condenser is used.
    (ii) For cut-backs, emulsions: water condenser is used.

Useful as a check on quality control and as a means of identifying volatile distillates and non-volatile residues.

  (b) *Water content test.* Can be obtained from distillation test or by use of a direct method (Dean-and-Stark method, in which a carrier-liquid is used in the distillation process). The sample is heated under reflux with an organic liquid which is immiscible with water. The carrier liquid distils into a graduated receiver carrying with it the water, which then separates to form the lower layer, the excess carrier lip overflowing from the trap and returning to the still. See BS 4385 for details.

  (c) *Loss on heating test.* The sample (50 g) is heated for a period (5 hr.) at temperature 163 °C in a ventilated oven. The weight loss is expressed as a percentage of original weight. See BS 4707 for details.

  (d) *Ash content test.* The sample (W g) is heated gently in a crucible until it begins to burn, and then ignited in a muffle furnace (at 775 °C) until the ash is free from carbon. Then, percentage ash content $= w/W \times 100$, where w is the mass of ash in grammes. See BS 4450 for details.

  (e) *Solubility tests.* The sample (2 g) is dissolved in a given amount of solvent (*e.g.* carbon disulphide for bitumens and toluene for tar), followed by filtration. The residue is determined quantitatively and the percentage soluble material deduced by difference. See BS 4690 for details.

PREMIXED BITUMINOUS MATERIALS

9. Introduction. Premixed bituminous materials consist of:

(a) *Mineral aggregate* (*e.g.* crushed rock, gravel, sand, limestone, blastfurnace slag) which is classified according to size into:

> (i) *Coarse* aggregate (aggregate retained on the 3 mm (⅛ in.) BS sieve).
> (ii) *Fine* aggregate (aggregate (including filler) passing the 3 mm (⅛ in.) BS sieve).
> (iii) *Filler* The portion of the fine aggregate which passes the No. 200 BS sieve (75 microns).

(b) *Binder:* bitumen (cut-back or penetration-grade and/or tar.

Two main types of premixed bituminous materials can be distinguished:

(a) *The "mastic" type* which contains high binder content of high viscosity with mineral aggregates acting as extender/filler. This type is impermeable (no voids) and therefore very durable, but expensive. Its strength and stability are mainly dependent on the binder type and content.

(b) *The "coated chippings" type* which contains low binder content of low viscosity with a relatively high proportion of mineral aggregates. It is very porous and therefore of low durability, but relatively inexpensive. Its strength and stability are mainly dependent on the aggregate-to-aggregate contact (interlock and friction).

Mixes ranging between the "mastic" and "coated chippings" types are used in practice. The properties of the mix are governed by its composition (binder: aggregate ratio).

10. Types of bituminous materials mixes.

(a) Mastic asphalt (BSS 1446, 1447).
(b) Rolled asphalt (BS 594).
(c) Asphaltic concrete.
(d) Macadam:

(i) Bitumen macadam: close-textured or dense (about 35 per cent fines); open- and medium-textured (BSS 1621,

2040); open (15 per cent fines or less); and medium (about 25 per cent fines).

(*ii*) Tar macadam (BSS 802, 1241, 1242): open- and medium-textured or dense.

(*e*) Cold asphalt (BS 1690).

(*f*) Dense tar surfacing (no BS yet).

(*a*) *Mastic asphalt* (BSS 1446, 1447). A type of asphalt composed of suitably graded mineral matter and asphaltic cement in such proportions as to form a coherent, voidless, impermeable mass, solid or semi-solid under normal temperature conditions, but sufficiently fluid when brought to a suitable temperature to be spread by means of a float.

It is normally heated to about 200 °C and is laid on a rolled asphalt basecourse or directly on a concrete base. Roughened surfaces can be obtained with surface dressing or chippings to improve skid resistance, the thickness of the finished surface being 25–50 mm.

Mastic asphalt is extremely durable and waterproof, very resistant to deformation and to softening by oil droppings (hence suitable for use at bus stops).

Normally used in roadwork, especially as a wearing course material in the construction of flexible road. Other uses include damp proof courses, roofing, flooring and tanking.

(*b*) *Rolled asphalt* (BS 594). This is defined (BS 892) as a material used as a dense wearing course, basecourse or roadbase material. It consists of a mixture of aggregate and asphaltic cement. (Asphaltic cement is defined as bitumen, a mixture of lake asphalt and bitumen, or lake asphalt and flux oils or pitch or bitumen, having cementing qualities suitable for the manufacture of asphalt pavements.)

The aggregate is often a mixture of sand and broken stone, slag or clinker. Rolled asphalt is normally laid hot at temperatures 110–190 °C, depending on the type of aggregate used and is then consolidated to a depth of 38–76 mm using a six-ton roller. Rolled asphalts are defined by their coarse aggregate content, *i.e.* the percentage in the total mixture of aggregate retained on the No. 7 BS sieve (2·4 mm). The low coarse aggregate content mixtures (containing about 30 per cent) are used solely for wearing courses and normally have coated chippings rolled into the surface. The high coarse aggregate content mixtures (containing about 60 per cent) are usually limited to basecourses and roadbases, although they may also be used for wearing courses.

Rolled asphalt is characterised by its durability, imperviousness and good load-spreading properties. It is relatively inexpensive and is quick and easy to lay with minimum inconvenience to road users. Its working life is at least twenty years. Its stability depends partly on the interlocking of the aggregates but mainly on the bitumen binder.

It is a suitable material for use in roadwork and is mostly used for heavy-duty highways, turnpikes and airfields.

(c) *Asphaltic concrete*. Frequently but erroneously known as "Marshall Asphalt" in the U.K. It is composed of a mixture of aggregate, sand and bitumen, the bitumen content being lower than in rolled asphalt. Its stability depends partly on the bitumen binder and partly on the aggregates (continuous-graded crushed aggregate) as compared with the stability of the mastic asphalt, which depends entirely on the bitumen binder.

It is used in heavy-duty dense road surfacing.

(d) *Macadam*.

(i) *Bitumen macadam* (BSS 1621, 2040). This is coated macadam in which the binder is wholly or substantially bitumen (BS 892).

Coated macadam is defined as "a road material consisting of graded aggregate that has been coated with a tar or bitumen, or a mixture of the two and in which the intimate interlocking of the aggregate particles is a major factor in the strength of the compacted roadbase or surfacing."

The binder may be cut-back bitumen or penetration-grade bitumen softer than 100 penetration. The graded aggregate may consist of crushed rock, slag or gravel.

The properties of bitumen macadam depend on the choice of aggregates (type and grading) and the binder (type, viscosity and content). It is cheap and easy to lay, requiring medium temperature only. Two main types can be distinguished.

I. Dense or close-textured bitumen macadam, which is similar to asphaltic concrete but having higher void content and stability mainly dependent on the interlocking of aggregate particles.

It is suitable for normal main road surfacing and as roadbase material in the construction of flexible roads.

II. Open- and medium-textured bitumen macadam, which has still higher void content, good skid resistance and free-draining properties. It is used for road resurfacing.

(*ii*) *Tar macadam* (BSS 802, 1241, 1242). Similar to bitumen macadam but having tar instead of bitumen as the binder.

(*e*) *Cold asphalt* (BS 1690). A close-textured type (BS 892) of coated macadam wearing course material, consisting of aggregate wholly passing a 6 mm ($\frac{1}{4}$ in.) BS sieve for the fine grade, and wholly or substantially passing a 10 mm ($\frac{3}{8}$ in.) BS sieve for the coarse grade, coated with a binder solely or substantially of bitumen, the composition of the mixture being so adjusted that the material can be spread and compacted while cold or warm and, if required, after storage.

BS 1690 gives specifications for two grades: fine cold asphalt and coarse cold asphalt. The coarse grade is the more recent type and has been introduced to meet cases where a slight extra thickness is required or where it is desired to have a rougher surface texture without the application of coated chippings.

It is suitable for lightly-used roads. It can be laid easily and cheaply, even in damp weather.

(*f*) *Dense tar surfacing* (no BS yet). BS 892 defines it as "a hot-process wearing course material consisting of aggregate, filler and road tar, in such gradings and proportions that when spread and compacted it provides a close-textured impervious mixture".

It is laid and compacted while hot. Unlike coated macadam, its strength depends to a large extent on the mortar binder. Like asphalts, it is classified according to the stone content of the mixture: 35 per cent stone content with coated chippings rolled into the surface, and 50 per cent stone content, which provides a sufficiently rugous surface without the application of coated chippings.

Dense tar surfacing provides an impervious surface which is resistant to softening by oil droppings and provides improved resistance to skidding as a result of the gradual weathering of the tar binder on the surface.

Table XXXV gives some typical mixes.

TABLE XXXV: PRACTICAL BITUMEN–AGGREGATE MIXES

| Type of bitumen–aggregate mix | BINDER (BITUMEN) | | FILLER | |
	Typical grade (penetration at 25 °C)	Typical content (% W on mineral aggregate)	Typical content (% W on mineral aggregate)	Typical void content (% V)
Mastic asphalt	20–30	12·0	30	0
Rolled asphalt	40–70	8·0	9	3
Asphaltic concrete	60–100	6·5	9	4
Dense bitumen macadam	100–250	5·0	5	10
Open bitumen macadam	150–350	4·5	5	30
Cold asphalt	200–500	5·5	12	15

11. Mix properties. The bituminous materials mixes are designed, in the case of pavements, to resist the destructive effects of traffic and weathering.

The main properties are:

(a) *Stability:* resistance to deformation under load.

(b) *Flexibility:* resistance to cracking due to persistent loading, repeated cycles of flexing and volume changes.

(c) *Durability:* resistance to moisture and weathering.

(d) *Workability:* ease of working, spreading and compacting.

(e) *Safety:* resistance to skidding (for wearing courses).

(f) *Impermeability:* to prevent water from penetrating to the base.

(g) *Stiffness:* to provide an effective distribution of traffic loads (high modulus of elasticity required).

Table XXXVI shows how each main property may be best achieved by suitable choice of materials and proportions.

12. Choice of mix. Mixes are designed on the basis of high durability and low cost. Other important factors to be considered include service conditions, types of aggregates available and cost of labour, equipment and plant.

(a) *Mixing.* This is carried out in the "Asphalt Plant" or "Hot-mix plant" which works mainly on the batch or discontinuous principle. Continuous mixers are also available.

(b) *Placing.* The mix can be placed or laid by hand or by machine, depending on the type or size of job. The base or sub-base must be sound and properly prepared before laying the materials.

TABLE XXXVI: EFFECTS OF COMPOSITION ON
MIX PROPERTIES

Mix property	Choice of:		
	Aggregate	Bitumen	
		Grade	Content
Stability	dense gradings, harsh texture	hard	low
Flexibility	open grading, or	soft	high
	dense grading	hard	high
Durability	dense grading	—	high
Workability	rounded	soft	high
Safety:			
(a) skid resistance	harsh	—	low
(b) fretting and ravelling	—	soft	high
Impermeability	dense grading	—	high
Stiffness	dense grading	hard	low

(c) *Compacting.* The purpose of compacting is to ensure that there is sufficient interlock between the aggregate particles to resist the action of the traffic. Compacting is done by rollers which can be one of three types:

(i) dead-weight, three-wheel and tandem;
(ii) vibrating;
(iii) pneumatic-tyred.

STRUCTURAL LAYERS IN A MODERN FLEXIBLE ROAD

The terminology for these structural layers is given in BS 892 and shown in Fig. 28.

13. Wearing course. Purpose:

(a) to provide an even surface which is skid-resistant and durable;
(b) to withstand the destructive forces of traffic loads and weathering.

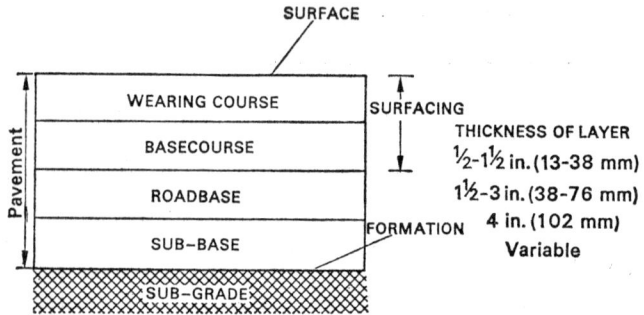

FIG. 28.—*Structural layers in a modern flexible road.*

Materials used:

     (*i*) Mastic asphalt.
     (*ii*) Rolled asphalt.
     (*iii*) Dense tar surfacing.
     (*iv*) Open-textured bitumen macadam.
     (*v*) Dense bitumen macadam.
     (*vi*) Cold asphalt.
     (*vii*) Tar macadam.

14. Basecourse. Purpose:

(*a*) to provide a surface for the laying of the wearing course;

(*b*) to distribute the traffic loads evenly over the roadbase.

Materials used:

     (*i*) Rolled asphalt.
     (*ii*) Dense bitumen macadam.
     (*iii*) Dense tar macadam.

15. Roadbase. Purpose:

(*a*) to provide the main load-bearing or distributing layer in the road structure.

Materials used:

     (*i*) Dense tar macadam.
     (*ii*) Dense bitumen macadam.
     (*iii*) Rolled asphalt.
     (*iv*) Dry-bound macadam.

(v) Wet-mix macadam.
(vi) Soil cement.
(vii) Cement-bound granular base.
(viii) Lean concrete.

16. Sub-base. Purpose:

(a) to provide a further and final distribution of loads on to the sub-grade;
(b) to provide an adequate thickness of frost-resistant material;
(c) to provide a working platform on which to lay the main construction.

Materials used:

(i) Granular material, *e.g.* crushed rock, slag, natural sand, gravel.
(ii) Soil cement.

17. Sub-grade. Should be adequately sound for the laying of the sub-base and sufficiently protected against possible defects in the road.

APPLICATIONS OF BITUMINOUS MATERIALS

A wide and varied range of applications exist.

18. Asphalt. Use: as finish for flat roof construction.

BS 988—Mastic asphalt for building (limestone aggregate).
BS 1162—Mastic asphalt for building (natural rock asphalt aggregate).
CP 144—Part 2—Roof coverings.

19. Bituminous felt. Use: as finish for roofs. Also used for damp-proof courses (DPC) and flashings. Three main types according to nature of base.

Type 1—Fibre base, with various finishes.
Type 2—Asbestos base, with various finishes.
Type 3—Glass fibre base, with various finishes.

BS 747—Roofing felts.
CPP 144, 101—Roof coverings.

20. Flooring and tiles.

(a) Mastic asphalt.

BS 1076—Mastic asphalt for building (limestone aggregate).
BS 1410—Mastic asphalt for building (natural rock asphalt aggregate).
BS 1451—Coloured mastic asphalt building (limestone aggregate).

(b) Pitch mastic.

BS 1450—Black pitch mastic flooring.
BS 3672—Coloured pitch mastic flooring.

21. Sealants, adhesives and coatings.

(a) Bitumen (BS 3712).
(b) Bitumen-rubber (BS 3712).
(c) Bitumen-based adhesives.

BS 3940—Classification for adhesives based on bitumen or coal tar.

(a) Bitumen-based coatings for ferrous products (BS 4147).
(b) Black bitumen oil varnish (BS 3634).

22. Paving and roadwork.

(a) Stabilisation of soils (bituminous emulsions, cutbacks).
(b) Surface treatment (road surfaces—bituminous emulsions).

BS 434—Bitumen road emulsions (anionic and cationic).
BS 3690—Bitumens for road purposes.

23. Electrical insulation.

BS 1858—Bitumen-based filling compounds for electrical purposes.

24. Tanking and D.P.C.

BS 1097—Mastic asphalt for tanking and damp-proof courses (limestone aggregate).

BS 1418—Mastic asphalt for tanking and damp-proof courses (natural rock asphalt aggregate).

PROGRESS TEST 8

1. Distinguish between "bitumen" and "asphalt." **(1, 2)**

2. What are the sources of bitumen ? **(3)**

3. How can the viscosity of bitumen be reduced ? **(4)**

4. Differentiate between "cut-back bitumen" and "bitumen emulsions." **(4(a), (b))**

5. Define the terms "tar" and "pitch." Outline the production and main applications of road tars. **(pp. 136–7)**

6. Give an account of the various methods of test commonly used for the determination of the flow properties of bitumen or tar. **(5)**

7. What is the purpose or significance of the following methods of test:

(a) specific gravity, (b) flash and fire points, (c) distillation, (d) water content, (e) loss on heating, (f) solvent solubility ? **(6–8)**

8. What is the composition of premixed bituminous materials ? Give an outline account of the characteristic properties and main applications of the various types of bituminous materials mixes commonly used in practice. **(9–10)**

9. Show diagrammatically the structural layers in a modern flexible road and describe, with examples, the purpose of each layer. **(13–17)**

10. Outline the general applications of bituminous materials. **(18–24)**

EXAMINATION QUESTIONS

1. (a) Outline the production of bituminous binders from:

(i) crude coal tar, (ii) "topped" petroleum, (iii) Trinidad Lake Asphalt.

(b) What are the main properties of bituminous materials that are of value in the construction industries ?

(P.S.B. B.Sc. Bldg.)

2. Distinguish between *mastic asphalt* and *rolled asphalt* and describe their properties and uses.

Explain the meaning of penetration-grade bitumen, and describe the associated test method.

(P.S.B. B.Sc. S.E.)

3. (a) Differentiate between:

(i) bitumen and asphalt, (ii) tar and pitch, (iii) cut-back bitumen and bitumen emulsion.

(b) Discuss, with instrumental details, the method of test for:

(i) the viscosity, (ii) the consistency (penetration), and (iii) the softening point,

of a bituminous material.

Indicate the relation between these properties and show how the penetration index can be evaluated from the values of these properties.

(P.S.B. B.Sc. S.E.)

4. (a) Explain why the time factor is more significant in the testing of plastic materials than in the testing of metals.

(b) Describe one of the dynamic simulative tests used for bituminous road materials.

(P.S.B. B.Sc. Bldg.)

5. Compare the composition and characteristic properties of:

(i) mastic asphalt, (ii) rolled asphalt, (iii) asphaltic concrete, (iv) bitumen macadam.

Comment on the practical applications of *each* of them.

(P.S.B. B.Sc. S.E.)

6. (a) Differentiate between:

(i) bitumen and tar, (ii) bituminous emulsion and cut-back, (iii) mastic asphalt and coated chippings.

(b) Discuss the design of bitumen–aggregate mixes in road-work.

(S.O.E. Grad. Exam. C.E. Specimen)

7. Outline the main properties of bitumen–aggregate mixes for pavements and describe how these mix properties can be controlled by choice of aggregate and bitumen.

(P.S.B. B.Sc. S.E.)

FURTHER READING

Road Research Laboratory D.S.I.R., *Bituminous materials in road construction*, H.M.S.O., 1962.

Hatherly, L. W., and Leaver, P. C., *Asphaltic Road Materials*, Edward Arnold, 1967.

CHAPTER IX

POLYMERS

WHAT IS A POLYMER?

A polymer is a chemical substance made up of repeating units, each unit being called a *monomer*.

$$monomers \xrightarrow{\text{polymerisation}} polymer$$

Proteins (animal origin) and cellulose (plant origin) are natural polymers. However, this chapter will deal mainly with synthetic organic polymers (plastics and rubbers) and synthetic inorganic polymers (silicones).

POLYMERISATION

The process of linking the monomers together to form a polymer is called polymerisation. Two main types of polymerisation reaction can be distinguished—addition polymerisation and condensation polymerisation.

1. Addition polymerisation (chain reaction). This involves the straightforward addition of monomers of the same kind or of different kinds (copolymerisation):

$$A + A + \cdots \rightarrow A\text{–}A\text{–}A\text{–}A\text{–}\cdots \text{ (homologous type)}$$
$$e.g. \text{ polythylene}$$
$$\text{or } A + B + \cdots \rightarrow A\text{–}B\text{–}A\text{–}B\text{–}\cdots \text{ (copolymer type)}$$
$$e.g. \text{ vinylchloride-acetate copolymer}$$

The conditions required for addition polymerisation are usually temperature and pressure and the presence of a catalyst. The mechanism of reaction is complex and is a three-stage process which involves initiation of the monomer (M) to form a free radical ($M-*$), propagation into a linear chain

$(M-M-M-M-*)$, followed by termination of the chain by combination of the chains or by means of a chain stopper (X):

Initiation: $M \rightarrow M-*$
monomer   free radical

Propagation: $M-* \rightarrow M-M-M-M-*$

Termination: $M-M-M-M-*$ $*-M-M-M-M \rightarrow$
$M-M-M-M-M-M-M-M$,   or
$M-M-M-M-* + X \rightarrow M-M-M-M-X$

2. Condensation polymerisation (step reaction). This involves polymerisation reactions between two monomers (usually bifunctional) with the elimination of a simple by-product, such as water, hydrogen chloride, etc.:

$$A + B \rightarrow AB + \text{simple by-product}$$
(*e.g.* nylon, terylene)

The mechanism of polymerisation is a stepwise process, which can be illustrated by the reaction between an alcohol and a carboxylic acid:

dihydric alcohol (*e.g.* glycol)   carboxylic acid (*e.g.* adipic acid)
(bifunctional)   (bifunctional)

and so on

STRUCTURE OF POLYMERS

Addition polymerisation usually forms linear chain polymers with varying degree of branching. Condensation polymerisation gives rise to cross-linked or network polymers.

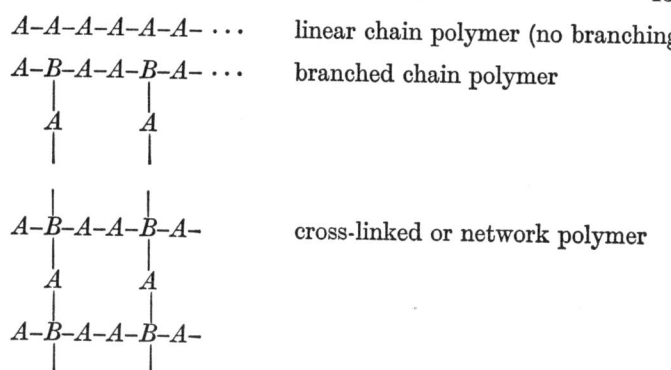

Linear and branched chain polymers are normally flexible and the flexibility decreases with increase in chain length and in the amount of branching. Cross-linked and network (three-dimensional) polymers are more rigid, the degree of rigidity increasing with the degree of cross-linking. Polymers with a wide range of properties—physical, mechanical and chemical—can therefore be obtained by controlling the degree and type of polymerisation.

CLASSIFICATION OF POLYMERS

It is convenient to classify polymers into two distinct classes—thermoplastics and thermosetting materials.

3. Thermoplastics. These belong to the linear and branched chain polymers which are obtained by addition or condensation polymerisation of bifunctional monomers. They can be softened and resoftened repeatedly by application of heat.

4. Thermosetting materials. These belong to the three-dimensional cross-linked or network polymers. They are the products of condensation polymerisation of monomers of functionality $\geqslant 3$. They will undergo setting and hardening on heating. They cannot be resoftened once they have set and hardened. Some common polymers are given on p. 156:

Thermoplastics	*Thermosetting*
Polyethylene	Phenol-formaldehyde (PF)
Polypropylene	Urea-formaldehyde (UF)
Polyvinylchloride (PVC)	Melamine-formaldehyde (MF)
Polystyrene	Polyesters (unsaturated)
Polytetrafluoroethylene (PTFE)	Epoxies
Nylon	Polyurethanes
Acrylics	
Polyesters (saturated)	

GENERAL PROPERTIES OF PLASTICS MATERIALS

Polymers (sometimes referred to as resins) are generally compounded with other materials (such as fillers, plasticisers, pigments, etc.) and the finished products are known as *plastics*.

5. Specific gravity. Plastics are generally light, specific gravity ranging from 0·9 (for polypropylene) to 3·2 (for rigid epoxies). Cellular plastics are even lighter (specific gravity about 0·1 or less). For comparison, SG of timber = 0·32–1·04; metals = 2·64–11·37; gravel concrete = 2·24–2·48.

6. Thermal properties.

(a) *Combustibility.* Plastics (having carbon backbone) are all combustible, although some of them (PVC, PTFE) do not support combustion. For most types of plastics the maximum service temperature is about 100 ± 50 °C.

(b) *Thermal conductivity.* Relatively low thermal conductivity and therefore good thermal insulating properties, particularly the cellular or expanded plastics (*see* Fig. 29).

(c) *Thermal expansion.* One of the main disadvantages is the relatively high thermal expansion of plastics (*see* Fig. 30).

(d) *Specific heat capacity.* This is the heat or energy necessary to raise the temperature of 1 kg of substance by 1 °K. Heat is essential in the manufacture of plastics material. The greater the specific heat capacity, the greater the amount of heat required and the greater the cost of the

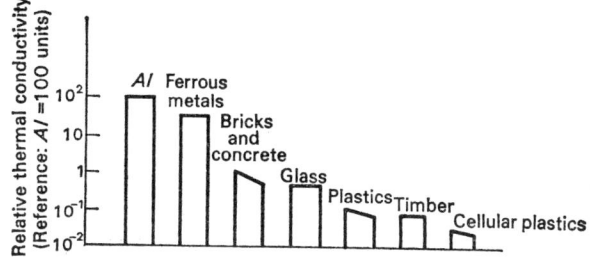

FIG. 29.—*Relative thermal conductivity of plastics materials.*

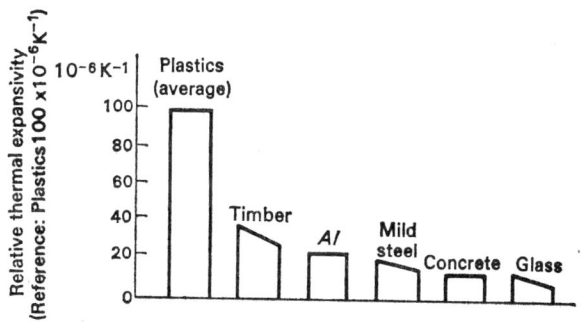

FIG. 30.—*Relative thermal expansivity of plastics materials.*

heating process. The values for specific heat capacities of plastics lie between the range 800–230 J kg^{-1} K^{-1}, whereas for brass and steel the value is about 400 J kg^{-1} K^{-1}.

7. Strength. Compared with metals and most building materials, plastics are relatively weak in tension as well as in compression. However, their strength/weight ratio is surprisingly high (*see* Figs. 31 and 32). However, strength can be improved by using reinforcing fillers such as glass fibre in glass-reinforced plastics.

8. Rigidity. Plastics have low rigidity compared with traditional building materials, except pinewood. Plastics

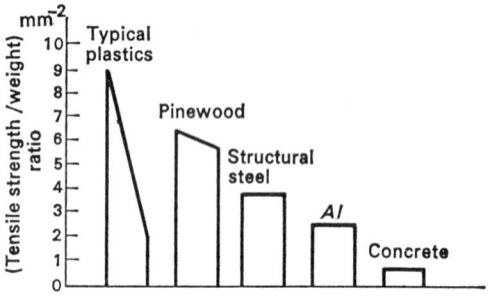

FIG. 31.—*Tensile strength/Weight ratio.*

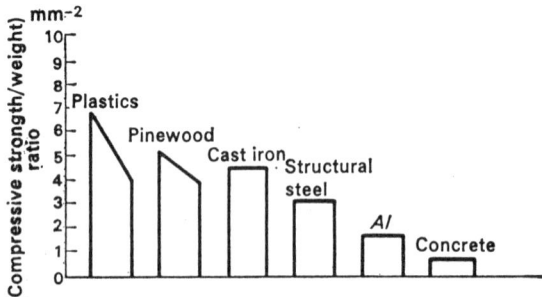

FIG. 32.—*Compressive strength/Weight ratio.*

materials have therefore a low modulus of elasticity which is a measure of rigidity or stiffness. Hence they are not suitable for use as conventional load-bearing materials (*see* Fig. 33). However, the modulus of elasticity can be improved by use of reinforcing fillers such as glass fibre in glass-reinforced plastics.

9. Durability.

(*a*) *Effects of sunlight.* Most plastics materials, if untreated, become brittle and yellow when exposed over a long period to ultra violet radiation from sunlight.

(*b*) *Corrosion resistance.* They are generally resistant to weathering, soils and usual service conditions. In this respect they are considered far superior to concrete, timber and most metals.

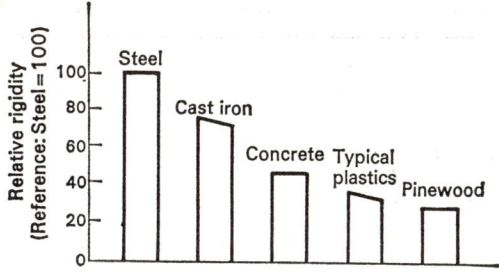

Fig. 33.—*Relative rigidity of plastics materials.*

10. Electrical properties. Plastics are good electrical insulators and hence are widely used in electrical engineering. One main disadvantage, however, is that they can acquire electrostatic charges (such as by friction) and may cause sparking which can be a fire hazard.

GENERAL APPLICATIONS OF PLASTICS MATERIALS

The ease of fabrication and the low cost of maintenance make plastics one of the most versatile materials used in everyday life. Some of the main applications of plastics in building are summarised in Table XXXVII.

THERMOPLASTICS

11. Polyethylene (polythene).

$$n\,CH_2 = CH_2 \xrightarrow[\text{polymerisation}]{\text{addition}} (-\,CH_2 - CH_2 -)_n$$

ethylene   polyethylene

Two types of product are obtained, depending on the condition of polymerisation:

(a) *High-density* (*HD*) *polyethylene*, which has a mainly linear structure with a high degree of crystallinity (ordered arrangement of polymers in bulk); this is obtained by a low-pressure process carried out catalytically in solution.

(b) *Low-density* (*LD*) *polyethylene*, which has a linear structure with considerable chain branching and lower

TABLE XXXVII: MAJOR APPLICATIONS OF PLASTICS IN BUILDING

Gutters, downpipes and fittings	Rigid PVC (polyvinylchloride)
Soil and waste pipes and fittings	Rigid PVC; LD (low-density) polythene; HD (high density) polythene; ABS (acrylonitrite butadiene styrene)
Water and gas pressure pipes and fittings	Rigid PVC, LD and HD polythene; ABS
Hot and cold water pipes	Chlorinated PVC; ABS; polypropylene; rigid PVC; polythene
Bathroom suites	GRP (glass-reinforced plastics); acrylics
Lavatory seats	Phenolics; aminos
Kitchen and bathroom, heart and pod units	GRP; acrylics
Cisterns—cold water storage	Rigid PVC; polystyrene: polythene
Roof lights and roofsheets	Rigid PVC (wire-reinforced and un-reinforced); GRP; acrylics
Vapour barriers, DPC and membranes	LD polythene film; PVC film
Door and other carpentry accessories	Nylon; rigid PVC; polyacetals; polyamides
Electrical conduit and fittings	Plasticised PVC; rigid PVC phenolics
Glazing other than rooflight	Acrylics
Heating system ducts	Rigid PVC; polypropylene
Insulation, including cold storage	Cellular plastics (polystyrene; PVC; polyurethane; phenolics and urea-formaldehyde)
Structural applications	GRP
Cladding	Rigid PVC; GRP; exterior quality melamine-surfaced phenolic laminates; acrylics; ABS; PVC-coated steel; PVF (polyvinylfluoride)-coated steel; various brush- and spray-applied plastics-based finishes
Decorative internal wall and ceiling linings	Melamine-surfaced phenolic laminates; paper- and fabric-backed PVC-based-film; unsupported vinyl film; PVC solution coatings; two-part polyester, epoxide and polyurethane paints; nylon fibre-reinforced paints; polystyrene; rigid PVC and expanded polystyrene tiles
Floorings	Rigid PVC; rigid and semi-rigid tiles, plasticised PVC-based tiles and sheets; *in situ* PVAc (polyvinylacetate); acrylic, polyester and epoxide resin mixed with Portland cement; PVC, epoxide, polyester and acrylic floor paints
Doors and windows	Rigid PVC; GRP
Plywood and particle board bonding	Phenolics and UF (Urea-formaldehyde resins

[The Construction Industry Handbook.

degree of crystallinity; this is obtained by a high-pressure process involving high temperature reaction in the vapour phase. Typical properties are given in Table XXXVIII.

TABLE XXXVIII: TYPICAL PROPERTIES OF POLYETHYLENE

	LD polyethylene	HD polyethylene
Density ($\times 10^3$ kg m^{-3})	0·91–0·93	0·94–0·97
Softening point (Vicat)	80 °C	90–100 °C
Thermal conductivity (W m^{-1} K^{-1})	0·13	0·42–0·55
Thermal expansivity ($\times 10^{-6}$ K^{-1})	120–140	120
Specific heat capacity (J kg^{-1} K^{-1})	2310	2100–2310
Tensile strength (MN m^{-2})	7–15	20–30
Compressive strength (MN m^{-2})	9–10	20–25
Elongation at break	300–700%	300–800%
Young's modulus (MN m^{-2})	120–240	550–1050
Effect of sunlight (UV)	Loss of strength and embrittlement	
Effect of chemicals	Attacked by strong acids only	
Optical properties	Translucent	Translucent–opaque
Electrical properties	Good electrical insulators	Good electrical insulators
Processing	Injection Blow moulding Extrusion Thermoforming	Blow moulding Extrusion Coating

Polyethylene has wide applications, because of its low cost, ease of processing, excellent electrical properties, excellent resistance to chemicals and moisture, odourless and non-toxic properties and satisfactory toughness and flexibility, even at low temperatures. The main disadvantages are a low softening point, high thermal expansivity, low tensile strength, high gas permeability and low rigidity.

The main uses of polyethylene (Table XXXVII) are in the form of sheets and films for packaging, membranes, temporary glazing and weather protection, and concrete curing; in the form of pipes and tanks for cold water storage tanks and cold water pipes; and in the form of mouldings for electrical insulation of wires and cables.

12. Polypropylene.

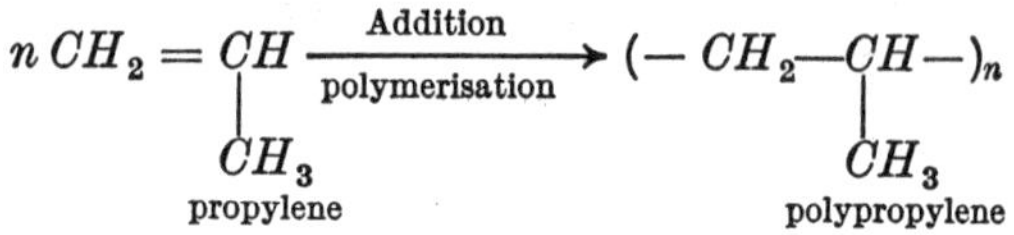

$$n\, CH_2 = CH \xrightarrow[\text{polymerisation}]{\text{Addition}} (-\,CH_2 - CH -)_n$$
$$\underset{\substack{|\\CH_3\\ \text{propylene}}}{} \qquad\qquad \underset{\substack{|\\CH_3\\ \text{polypropylene}}}{}$$

Three types of polymer may be obtained from polypropylene. These are known as:

(a) *isotactic*, when the branched groups (CH_3-) are all arranged on the same side of the chain,

(b) *syndiotactic*, with the branched groups (CH_3-) arranged on alternate sides of the chain, and

(c) *atactic*, with a random arrangement of the branched groups (CH_3-), as illustrated in Fig. 34.

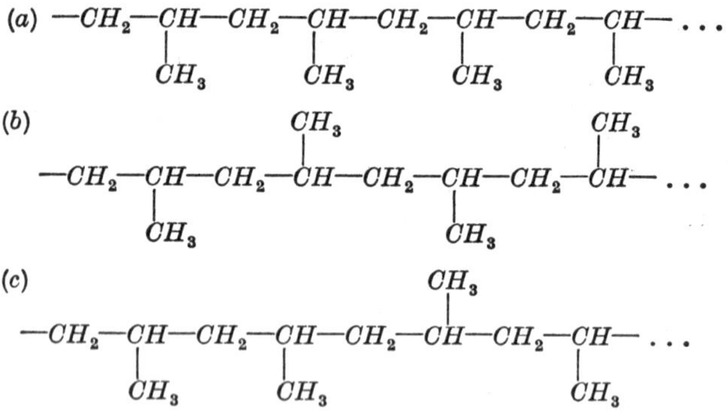

FIG. 34.—*Stereopolymers.*

(a) *Branched group (CH_3) arranged on same side of the chain: isotactic.*

(b) *Branched group (CH_3) arranged on alternate sides of the chain: syndiotactic.*

(c) *Random arrangement of branched group (CH_3): atactic.*

Typical properties of propylene are given in Table XXXIX. Polypropylene compares favourably with ethylene—lower density, higher softening point and hence higher service temperatures, better strength and rigidity, better resistance to chemicals and solvents. The mechanical properties can be further improved by fillers (*e.g.* glass fibre reinforcements).

Propylene finds wide application in the form of films and

TABLE XXXIX: TYPICAL PROPERTIES OF PROPYLENE

	Conventional	Glass fibre-reinforced (30%)
Density ($\times 10^3$ kg m^{-3})	0·90	1·10–1·16
Softening point (Vicat)	100–120 °C	112–120 °C
Thermal conductivity (W m^{-1} K^{-1})	0·088	—
Thermal expansivity ($\times 10^{-6}$ K^{-1})	120	54–85
Specific heat capacity (J kg^{-1} K^{-1})	1932	3570
Tensile strength (MN m^{-2})	33–35	34–54
Compressive strength (MN m^{-2})	35	40–60
Elongation at break	20–300%	5–20%
Young's modulus (MN m^{-2})	900–1400	Improved, depending on glass-fibre content
Effect of sunlight (*UV*)	Good in specially formulated or black grade	
Effect of chemicals	Attacked by strong acids	
Optical properties	Translucent	Opaque
Electrical properties	Good electrical insulators	
Processing	Injection Extrusion Blow moulding Thermoforming	Injection Extrusion

sheets, pipes, rods and profiles, filament and fibres, moulded articles (*e.g.* tanks, cisterns) and electrical insulators and connectors. A novel application is for hinged articles, because of its high flexibility and fatigue-resisting properties.

Copolymers of propylene with other monomers (*e.g.* ethylene) are possible.

13. Polyvinylchloride (PVC).

$$n\,CH_2 = CH \xrightarrow[\text{polymerisation}]{\text{addition}} (-\,CH_2\!-\!CH-)_n$$
$$|\phantom{\xrightarrow[\text{polymerisation}]{\text{addition}} (-\,CH_2\!-\!CH-)_n}|$$
$$Cl\phantom{\xrightarrow[\text{polymerisation}]{\text{addition}} (-\,CH_2\!-\!CH-)_n}Cl$$

vinyl chloride   polyvinylchloride

The polymerisation is carried out in presence of heat, pressure and catalyst (benzoyl peroxide) to give an atactic polymer (Fig. 34 (c)). Two different types are available:

(a) Rigid or unplasticised PVC, and
(b) flexible or plasticised PVC.

The typical properties are given in Table XL.

TABLE XL: TYPICAL PROPERTIES OF PVC

	Rigid PVC	Flexible PVC
Density ($\times 10^3$ kg m^{-3})	1·35	1·16–1·35
Softening point (Vicat)	56–85 °C	$\leqslant$79 °C
Thermal conductivity (W m^{-1} K^{-1})	0·16–0·29	0·13–0·17
Thermal expansivity ($\times 10^{-6}$ K^{-1})	50–60	70–250
Specific heat capacity (J kg^{-1} K^{-1})	840–2520	1260–2100
Tensile strength (MN m^{-2})	40	10–25
Compressive strength (MN m^{-2})	90	7–12
Elongation at break	60%	200–450%
Young's modulus (MN m^{-2})	2000–2800	3500–4800
Effect of sunlight (UV)	Tendency to decompose, except when suitably stabilised	
Effect of chemicals	Attacked by aromatics, soluble in ketones and esters	
	Not attacked by acids and alkalis	
Optical properties	Transparent to opaque	
Electrical properties	Good electrical insulators	
Processing	Extrusion	Extrusion
	Blow moulding	Blow moulding
	Injection	Injection
		Coating

PVC finds wide application in pipes, gutters and fittings, flooring and ceiling panels, electrical conduit and electrical cable insulation, roof sheets, cladding. It is not suitable for hot water pipes owing to its low softening point.

PVC (plasticised or unplasticised) can be produced in a cellular or expanded form, which is used for thermal insulating purposes (BS 3869), with a reduction in density and strength. Expanded rigid PVC is less inflammable, but more expensive, than expanded polystyrene.

Copolymers of vinylchloride with other monomers (such as vinyl acetate and acrylonitrile) are possible.

Copolymers of vinylchloride–vinylacetate have a low softening point and are mainly used in flooring.

Polyvinylacetate (PVAc), obtained from polymerisation of vinylacetate, is too soft to be used in moulded plastics. It has a density of $1 \cdot 19 \times 10^3 \, kg \, m^{-3}$ and atactic polymers are available commercially. PVAc is widely used in the building industry as emulsion paints, adhesives (for timber, expanded polystyrene, PVC wall coverings), bonding agents (for plaster etc.).

14. Polystyrene.

Polymerisation, carried out using a peroxide catalyst, gives the atactic and amorphous polystyrene. This is hard and brittle and has the characteristic metallic ring when struck. To reduce brittleness, polystyrene can be toughened by mixing with another polymeric material (usually styrene–butadiene copolymer), or by copolymerisation with acrylonitrile (to give SAN plastics) or acrylonitrile–butadiene (to give ABS plastics).

Typical properties of polystyrene and copolymers are given in Table XLI.

The main uses of polystyrene (*see also* Table XXXVII) are sheets and films (packaging), moulded articles (refrigerator parts, food containers), electrical insulation, light fittings, etc.

Expanded polystyrene is available in the form of tiles and boards and is used for insulation of walls, floors and ceilings (BSS 3832, 3837).

(*a*) *ABS.* This copolymer is extremely tough and strong, has good impact strength at low temperature, but is rather

TABLE XLI: TYPICAL PROPERTIES OF POLYSTYRENE AND COPOLYMERS

	Conventional	Toughened	SAN	ABS
Density ($\times 10^3$ kg m^{-3})	1·04–1·11	0·98–1·10	1·07–1·10	0·99–1·10
Softening point (Vicat)	82–103°C	78–100°C	85–103°C	85°C
Thermal conductivity (Wm^{-1} K^{-1})	0·09–0·21	0·04–0·17	0·09–0·21	0·04–0·30
Thermal expansivity ($\times 10^{-6}$ K^{-1})	60–80	34–210	60–80	60–130
Specific heat capacity (J kg^{-1} K^{-1})	1340–1466	1340–1466	1340–1425	1383–1676
Tensile strength (MN m^{-2})	35–62	17–45	55–83	17–58
Compressive strength (MN m^{-2})	89–110	28–62	96–117	17–76
Elongation at break	1–3%	8–50%	2–4%	10–140%
Young's modulus (MN m^{-2})	2410–4130	1720–3100	2756–2790	1378–3445
Effect of sunlight (*UV*)	Tendency to lose strength and to yellow slightly			
Effect of chemicals	Attacked only by strong oxidising acids			
Optical properties	Transparent	Translucent	Transparent	Translucent
Electrical properties	Good electrical insulators			
Processing	Injection Extrusion Thermo-forming Blow moulding	Injection Extrusion Thermo-forming Blow moulding	Injection Extrusion Blow moulding	Injection Extrusion Thermo-forming

expensive. It has a higher distortion temperature than PVC. It is resistant to acids and alkalis, being attacked only by strong oxidising acids. It is however soluble in esters, chlorinated hydrocarbons and ketones.

Its main uses include vacuum-formed baths, water pipes, sheets for cladding panels.

Cellular ABS is also available, and is used for insulation purposes (wall panel application, sandwich construction, infill for cavity).

(*b*) *SAN.* Similar to ABS. Compared to polystyrene, it shows improved tensile strength, hardness and rigidity. It is more widely used in Europe and in the U.S.A.

15. Polytetrafluoroethylene (PTFE) or "Teflon".

$$n \, CF_2 = CF_2 \xrightarrow[\text{polymerisation}]{\text{addition}} (-CF_2-CF_2-)_n$$

tetrafluoroethylene   polytetrafluoroethylene

Polymerisation is carried out using heat, pressure and catalyst (organic peroxide or oxygen). The product (PTFE) is a highly crystalline polymer which loses its crystallinity above 327 °C. It has excellent resistance to most chemicals and solvents. It has a very low coefficient of friction, as low as 0·04 (polyethylene: 0·25, PVC: 0·4) non-stick properties and good heat resistance.

PTFE is mainly used as moulding in aircraft, seals, gaskets, laboratory equipment; electrical insulation; for lining chutes and coating metal objects; in low friction bearings.

Typical properties are given in Table XLII.

TABLE XLII: TYPICAL PROPERTIES OF PTFE

Density ($\times 10^3$ kg m^{-3})	2·14–2·20
Softening point (Vicat)	260 °C
Thermal conductivity (Wm^{-1} K^{-1})	0·25
Thermal expansivity ($\times 10^{-6}$ K^{-1})	100
Specific heat capacity (J kg^{-1} K^{-1})	1050
Tensile strength (MN m^{-2})	15–35
Compressive strength (MN m^{-2})	10–15
Elongation at break	200–400%
Young's modulus (MN m^{-2})	350–620
Effect of sunlight (UV)	None
Effect of chemicals	Unaffected
Optical properties	Translucent–opaque
Electrical properties	Good electrical insulators
Processing	Injection, extrusion

16. Acrylics. The best-known polymer in this group is polymethylmethacrylate (PMMA), sold under the trade name of "Perspex". It is obtained by polymerisation of the monomer, methylmethacrylate, in presence of a peroxide catalyst.

$$n \, CH_2 = \overset{\displaystyle CH_3}{\underset{\displaystyle COOCH_3}{C}} \xrightarrow[\text{polymerisation}]{\text{addition}} (-CH_2-\overset{\displaystyle CH_3}{\underset{\displaystyle COOCH_3}{C}}-)_n$$

methylmethacrylate   polymethylmethacrylate

The resulting polymer is atactic and amorphous. PMMA is characterised by its high degree of optical clarity transmitting about 92 per cent of incident light, whereas glass has about 90 per cent light transmission. It is highly resistant to weathering, tougher than polystyrene but attacked by many chemicals.

Typical properties are given in Table XLIII.

Main uses of PMMA include sheeting, rooflights and domes, light fittings, display signs, aircraft glazing, sinks, baths (BS 4305), hospital equipment and sanitary ware. It is used also as an adhesive.

TABLE XLIII: TYPICAL PROPERTIES OF PMMA

Density ($\times 10^3$ kg m^{-3})	1·10–1·20
Softening point (Vicat)	71–91 °C
Thermal conductivity (Wm^{-1} K^{-1})	0·17–0·25
Thermal expansivity ($\times 10^{-6}$ K^{-1})	50–90
Specific heat capacity (J kg^{-1} K^{-1})	1470
Tensile strength (MN m^{-2})	55–75
Compressive strength (MN m^{-2})	80–130
Elongation at break	2·7%
Young's modulus (MN m^{-2})	2700–3500
Effect of sunlight (UV)	Very slight (appearance unaffected)
Effect of chemicals	Attacked by concentrated acids, but not by alkalis
Optical properties	Transparent (also with varying degrees of opacity)
Electrical properties	Good electrical insulators
Processing	Injection, extrusion, thermoforming, blow moulding

17. Polyamides and polyesters. These are obtained by condensation polymerisation:

(a) Amine + carboxylic acid → polyamide.
(b) Alcohol + carboxylic acid → polyester.

(a) *Polyamides.* The best-known linear polyamides (thermoplastics) are the nylons, which are obtained by condensation between a diamine and a dicarboxylic acid, or by direct polymerisation of amino acids:

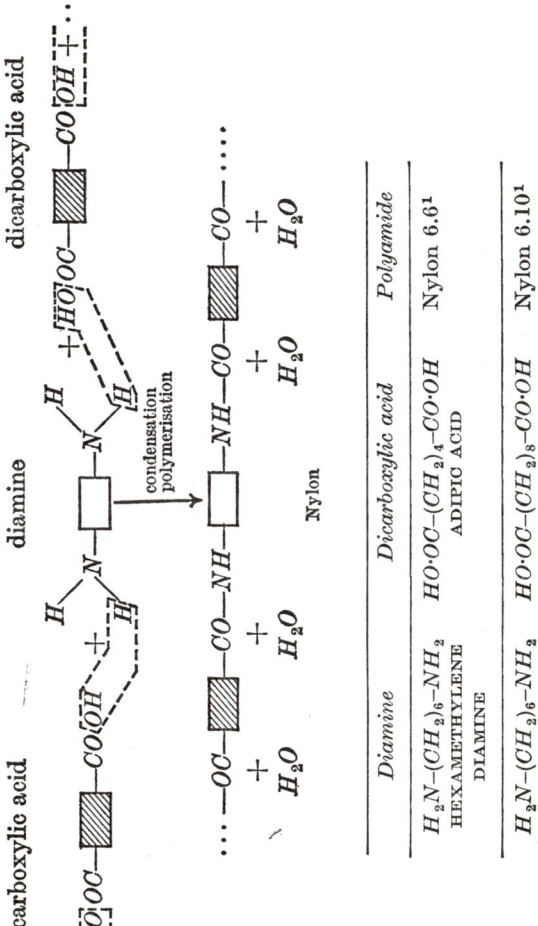

Diamine	Dicarboxylic acid	Polyamide
$H_2N-(CH_2)_6-NH_2$ HEXAMETHYLENE DIAMINE	$HO\cdot OC-(CH_2)_4-CO\cdot OH$ ADIPIC ACID	Nylon 6.6[1]
$H_2N-(CH_2)_6-NH_2$	$HO\cdot OC-(CH_2)_8-CO\cdot OH$ SEBACIC ACID	Nylon 6.10[1]

[1] The figures indicate the number of carbon atoms in the diamine and the dicarboxylic acid respectively.

170   PROPERTIES OF MATERIALS

The types of nylons in common use are 6.6; 6; 6.10; 11 and 12 (the latter being comparatively recent).

The important characteristics of nylons are high mechanical strength, resistance to wear and tear, low coefficient of friction (0·1–0·3) and comparatively high service temperatures. The disadvantages are, however, high cost and high moisture absorption, which affects dimensional stability and electrical insulation properties.

Glass-filled nylons are also available; they have improved strength and heat-distortion temperature.

Typical properties of nylons are given in Table XLIV.

Nylons find very wide applications:

>*Engineering:* gears, bearings, bushes, slides, metal coatings.
>*Transport:* automobile and car components.
>*Textiles:* fabrics.
>*Domestic appliances:* curtain rail fittings, door hinges, handles and latches, parts for washing machines, vacuum cleaners, valves for water supply.
>*Food industry:* packaging.
>*Electrical industry:* coil formers, computer components, power tools.

TABLE XLIV: TYPICAL PROPERTIES OF NYLONS

	Nylon 6.6	Nylon 6.6 +30% glass	Nylon 6.10	Nylon 6	Nylon 11
Density ($\times 10^3$ kg m^{-3})	1·14	1·39	1·09	1·13	1·04
Melting point	250–265°C	265°C	210–220°C	220–225°C	186°C
Thermal conductivity (Wm^{-1} K^{-1})	0·24	0·23	0·22	0·24	0·28
Thermal expansivity ($\times 10^{-6}$ K^{-1})	80	28	90	83	150
Specific heat capacity (J kg^{-1} K^{-1})	1675	1250	1250–2100	1675	2420
Tensile strength (MN m^{-2})	86	172	58·5	76	54
Compressive strength (MN m^{-2})	110	200	68	96	55
Elongation at break	40–80%	3%	50–200%	50–200%	180–380%
Young's modulus (MN m^{-2})	1200–2860	7850	1100–1930	760–3100	1270
Effect of sunlight (UV)	Affected	Slight	Affected	Affected	Affected
Effect of chemicals	Attacked by strong mineral acids only				
Optical properties	Translucent–opaque	Translucent	Translucent–opaque		Translucent
Electrical properties	Good electrical insulators				
Processing	Injection, Extrusion, Blow moulding	Injection, (see 32)	Injection, Extrusion, Blow moulding	Injection, Extrusion, Coating, Blow moulding	Injection, Extrusion, Blow moulding

(*b*) *Polyethylene terephthalate* (terylene). This linear (thermoplastic) polyester is obtained by condensation polymerisation of ethylene glycol and terephthalic acid. The resulting polymer has the structure:

$$\left(O{\cdot}OC-\left\langle \right\rangle-CO-O-CH_2-CH_2- \right)_n$$

It is usually available in the form of fibres (used in textiles, etc.), in film form (used in packaging and also in many electrical applications).

Typical properties of films:

Density: $1{\cdot}39 \times 10^3$ kg m^{-3}
Elongation at break: 50–130 per cent
Tensile strength: 120–170 MN m^{-2}.

18. Polyacetals. These are crystalline polymers of formaldehyde, which have the $-C-O-C-$ backbone.

Homopolymer
(acetal polymer)

Copolymer
(acetyl copolymer)

They have high strength and high rigidity which can be retained over a wide range of temperatures, low moisture absorption (hence good dimensional stability), low coefficient of friction, good resistance to high temperatures, chemicals and solvents and good electrical insulation. The copolymers have improved properties but they are usually more expensive.

They are used in various applications—car components,

door handles, bearings, components for domestic appliances, plumbing fittings, hot water piping, components for textile machinery, etc.

Typical properties are given in Table XLV.

TABLE XLV: TYPICAL PROPERTIES OF POLYACETALS

	Acetal polymer	Acetal copolymer
Density ($\times 10^3$ kg m^{-3})	1·425	1·410
Melting point	175 °C	163 °C
Thermal conductivity (Wm^{-1} K^{-1})	0·21–0·23	0·23
Thermal expansivity ($\times 10^{-6}$ K^{-1})	81	97
Specific heat capacity (J kg^{-1} K^{-1})	1460	
Tensile strength (MN m^{-2})	69	59
Compressive strength (MN m^{-2})	110–124	
Elongation at break	15–75%	23–35%
Young's modulus (MN m^{-2})	2750–3100	2750
Effect of sunlight (UV)	Slightly affected, unless stabilised	
Effect of chemicals	Attacked by strong acids and alkalis	
Optical properties	Transparent–opaque	
Electrical properties	Good electrical insulators	
Processing	Injection, blow moulding, extrusion	

19. Polycarbonates. This class can be considered as the polyesters of carbonic acid H_2CO_3. One important type has the structure:

They are expensive, but have remarkable properties such as exceptional dimensional stability, high impact strength, high distortion temperature (135–150 °C) and good mechanical properties which can be maintained over a wide range of temperature, good electrical properties and good chemical and weathering resistance. Polycarbonates, like polyacetals, have replaced metals in various applications. Mechanical properties are improved in glass-reinforced polycarbonates.

Main applications are:

> *Electrical applications:* light fittings, electrical fixtures, lamps.
> *Domestic appliances:* double insulation, table ware, fans.
> *Car fittings:* crash helmets, car components (*e.g.* light fixtures and horn rings).

Typical properties are given in Table XLVI.

TABLE XLVI: TYPICAL PROPERTIES OF POLYCARBONATES

	Unfilled	*Glass-filled*
Density ($\times 10^3$ kg m^{-3})	1·2	1·4
Softening point (Vicat)	215–226 °C	
Thermal conductivity (Wm^{-1} K^{-1})	0·19	
Thermal expansivity ($\times 10^6$ K^{-1})	66	
Specific heat capacity (J kg^{-1} K^{-1})	1260	
Tensile strength (MN m^{-2})	62	117
Compressive strength (MN m^{-2})	85	
Elongation at break	80	3
Young's modulus (MN m^{-2})	2200	
Effect of sunlight (UV)	Yellows slightly	
Effect of chemicals	Attacked by strong acids and alkalis	
Optical properties	Transparent–opaque	Translucent–opaque
Electrical properties	Good electrical insulators	
Processing	Injection, extrusion	(*See* **32**)

20. Cellulosics. These are derived from natural polymer cellulose ($C_6H_{10}O_5$)$_n$, a major constituent of plants. The ones commonly used are cellulose nitrate, cellulose acetate, cellulose acetate butyrate and cellulose acetate propionate.

(a) *Cellulose nitrate* is obtained by nitration of cellulose with a mixture of nitric and sulphuric acids. It is hard, tough and resistant to dilute acids and alkalis, has a good water resistance, but it is highly inflammable and affected by heat and sunlight. It is used in paint finishes, and some moulded articles.

(b) *Cellulose acetate* is obtained from the reaction of cellulose with acetic anhydride. Similar in properties to cellulose nitrate, except that it is non-inflammable and poorly water-resistant. It is used in packaging, domestic equipment, containers, electrical parts and as binder in emulsion paints.

(c) *Cellulose acetate butyrate* (*CAB*). This copolymer shows improved resistance to water, petrol and oil, and better toughness. It is used in car parts, containers, packaging, radio and television parts, illuminated signs and also for coatings.

THERMOSETS

21. Phenol formaldehyde (PF) plastics. This is obtained by condensation polymerisation of phenol (P) and formaldehyde (F) in an acid or alkaline medium. The reaction is rather complex, leading to a highly cross-linked three-dimensional network polymer (*see* p. 175).

The resin (pure polymer) is hard, rigid and resistant to heat, most chemicals and solvents. It has good mechanical strength and electrical insulation. However it discolours on aging. It is often mixed with fillers (such as woodflour, asbestos, glass fibre) pigments and other additives to extend the range of properties (*see* Table XLVII) and applications (*see* Table XXXVII). The colour range is limited to black or brown.

Phenol formaldehyde is mainly used as mouldings—electrical parts, domestic appliances, lavatory seats; surface coatings—stoving lacquers; impregnants for paper and fabric laminates (formica); paints and adhesives; cellular or foamed products (for thermal insulation).

22. Amino–formaldehyde plastics. Amino compounds (such as urea and melamine) can also condense with formaldehyde

$$P + F + P \longrightarrow P' - F' - P'$$

TABLE XLVII: TYPICAL PROPERTIES OF PHENOL FORMALDEHYDE MOULDING MATERIALS

	Unfilled	Woodflour-filled	Asbestos-filled	Glass-filled
Density ($\times 10^3$ kg m^{-3})	1·25–1·3	1·32–1·45	1·6–1·85	1·5–2·0
Heat-distortion temperature	112–125 °C	125–175 °C	160–200 °C	110–125 °C
Thermal conductivity (Wm^{-1} K^{-1})	0·125–0·25	0·17–0·29	0·42–0·46	0·125–0·25
Thermal expansivity ($\times 10^{-6}$ K^{-1})	25–60	30–45	6–10	80–85
Specific heat capacity (J kg^{-1} K^{-1})	1550–1760	1460–1670	1170–1340	1460–1880
Tensile strength (MN m^{-2})	35–55	45–60	28–55	40–70
Compressive strength (MN m^{-2})	70–210	150–280	110–210	70–210
Elongation at break	1·0–1·5	0·4–0·8	0·1–0·2	<0·5
Young's modulus (MN m^{-2})	5200–7000	5500–8000	9000–11 500	7000–14 000
Effect of sunlight (UV)	Slightly affected		Some darkening	
Effect of chemicals	Unaffected by solvents, but attacked by acids and alkalis			
Optical properties	Transparent	Opaque	Opaque	Opaque
Electrical properties	Good electrical insulators			
Processing	Compression, injection, transfer moulding			

to form highly cross-linked polymers which are clear and water white. They can be compounded with fillers, pigments and other additives to form moulding materials with a wide range of colours.

Urea–formaldehyde (*UF*) *and melamine–formaldehyde* (*MF*)

TABLE XLVIII: TYPICAL PROPERTIES OF AMINO FORMALDEHYDE

PLASTICS

	UF (woodflour filled)	*UF* (cellulose-filled)	*MF* (cellulose-filled)
Density ($\times 10^3$ kg m^{-3})	1·4–1·5	1·5–1·55	1·5–1·55
Maximum service temperature	75 °C	75 °C	100 °C
Thermal conductivity (Wm^{-1} K^{-1})	0·25–0·38	0·29–0·42	0·29–0·42
Thermal expansivity ($\times 10^{-6}$ K^{-1})	35–45	35–45	35–45
Specific heat capacity (J kg^{-1} K^{-1})		1670	1670
Tensile strength (MN m^{-2})	38–56	48–76	48–83
Compressive strength (MN m^{-2})	70–110	70–110	70–110
Elongation at break	0·5–1·0	0·5–1·0	0·5–1·0
Young's modulus (MN m^{-2})	7000–12 000	7000–13 500	7000–10 500
Effect of sunlight (*UV*)	Slight colour change		
Effects of chemicals	Attacked by strong acids and alkalis		Attacked by strong acids only
Optical properties	Opaque	Translucent–opaque	
Electrical properties	Good electrical insulators		
Processing	Compression (injection)		

have general properties similar to *PF*. *MF* has better resistance to chemicals, heat and moisture, is harder and scratch-proof, has better electrical insulating properties but is much more expensive.

Their main uses include mouldings, surface coatings, paints, adhesives, cellulose or foamed products.

Typical properties are given in Table XLVIII.

23. Polyesters (unsaturated). These are the products of condensation polymerisation of an unsaturated dicarboxylic acid (*e.g.* maleic acid) and a dihydric alcohol (*e.g.* glycol), followed by curing with a cross-linking agent (*e.g.* styrene):

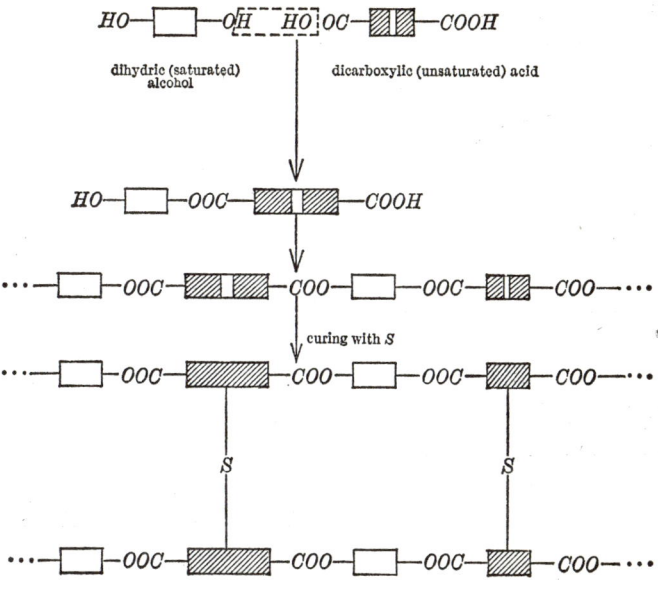

Thermoset polyester

The degree of cross-linking can be controlled by use of modifiers which are saturated dicarboxylic acids (*e.g.* phthalic acid). Consequently, a wide variety of polyester resins can be produced with a wide range of properties depending on the choice of monomers (alcohols and acids), modifiers and curing agents.

The resins are generally strong and rigid and have good resistance to heat and most chemicals except strong alkalis.

They are mainly used in paints, clear finishes, jointing, repair work and floorings.

They are also used as binders for glass fibres in glass-reinforced polyesters which are characterised by high impact strength, good dimensional stability, good resistance to heat, chemicals and weathering and good surface hardness. Glass-reinforced polyesters are used in building applications such as cladding panels, radomes, boat hulls, car bodies.

Typical properties are given in Table XLIX.

TABLE XLIX: TYPICAL PROPERTIES OF POLYESTERS

	Cured and unfilled	Glass fibre-filled
Density ($\times 10^3$ kg m^{-3})	1·1–1·4	1·8–2·3
Softening point (Vicat)	—	150–180 °C
Thermal conductivity (Wm^{-1}K^{-1})	0·17–0·19	0·42–0·67
Thermal expansivity ($\times 10^{-6}$K^{-1})	100–150	25–33
Specific heat capacity (J kg^{-1}K^{-1})	1260	1050
Tensile strength (MN m^{-2})	31–70	28–70
Compression strength (MN m^{-2})	90–240	150–200
Elongation at break	<5%	—
Young's modulus (MN m^{-2})	2800–7000	5500–20 000
Effect of sunlight (UV)	Affected, unless stabilised	
Effect of chemicals	Unaffected by solvents, attacked by strong alkalis	
Optical properties	Transparent–Opaque	Opaque
Electrical properties	Good electrical insulators	
Processing	Injection	Polymerisation (*see* **31**)

24. Epoxies. These are recognised by the highly-reactive epoxy groups which occur at each end of the molecular chain. They are produced by condensation polymerisation of epichlorhydrin with a polyhydroxy compound (*e.g.* bisphenol A).

Curing or cross-linking can take place through epoxide ($\triangle$) or hydroxyl (—OH) groups by cross-linking agents (*e.g.* amines, silicones, other resins, etc.). Cured epoxy resins are extremely tough and resistant to solvents and chemicals, heat, moisture and light. They have good adhesive properties (on most surfaces), and good electrical properties. However, they are expensive and have low softening temperature. Mainly used as adhesives ("Araldite"), electrical insulation, surface coatings and floorings and in repair work. Epoxy resins

Epichlorhydrin

bisphenol A

diepoxide (linear polymer)

are generally compounded with reinforcing fibres (such as glass fibres, carbon fibres) to enhance hardness and strength for structural applications similar to glass-reinforced polyesters.

Typical properties of epoxy materials are given in Table L.

TABLE L: SOME TYPICAL PROPERTIES OF EPOXY MATERIALS

	Rigid and unfilled	Flexible and unfilled	Glass fibre filled
Density ($\times 10^3$kg m^{-3})	1·2	1·2	2·0
Heat distortion temperature	Up to 300 °C	Up to 60 °C	150–250 °C
Thermal conductivity (Wm^{-1}K^{-1})	0·17–0·21	0·17–0·21	
Thermal expansion ($\times 10^{-6}$K^{-1})	50–90	50–90	
Specific heat capacity (J kg^{-1}K^{-1})	1250–1670	1250–1670	
Tensile strength (MN m^{-2})	35–85	70–280	42–105
Compressive strength (MN m^{-2})	105–210		105–210
Elongation at break	5–10%	10–100%	
Young's modulus (MN m^{-2})	1400–4200		2100
Effect of sunlight (UV)	Yellows slightly		
Effect of chemicals	Unaffected by chemicals, except chlorinated hydrocarbons and ketones		
Optical properties	Transparent–opaque		
Electrical properties	Good electrical insulators		
Processing	Compression, transfer, extrusion, blow moulding		Polymerisation (*see* 31)

(a) *Synthesis of polyesters and epoxies. General principles (Schematic):*

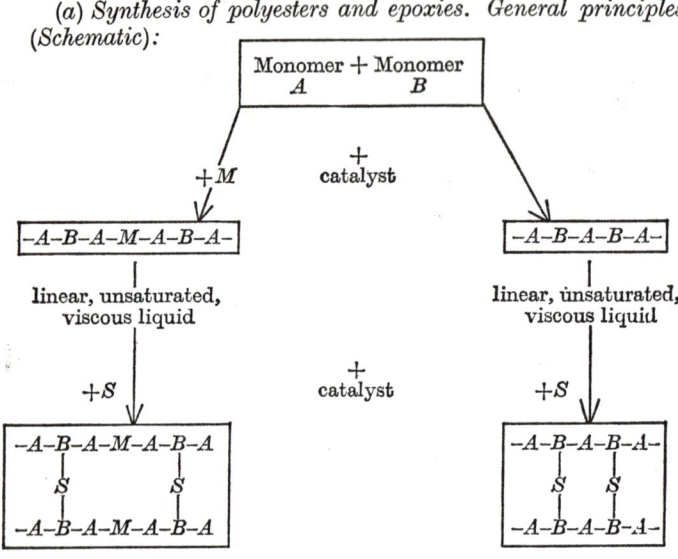

General principles of synthesis of polyesters and epoxies

	Polyesters	*Epoxies*
Reacting monomers		
A	Dihydric alcohol (glycols)	Polyhydric compound (bisphenol *A*)
B	Unsaturated acids (maleic acid)	Epichlorhydrin
Cross-linking monomers		
S	Styrene, Methylmethacrylate, Diallyl phthalate, . . .	Amines, Dicarboxylic acids, Isocyanates, . . .
Catalysts	Peroxides (benzoyl, MEK, HCH)	Tertiary amines
Modifiers		
M	Saturated dibasic acids (phthalic acids, HET)	Thiokol resins (liquid polysulphide rubbers)

MEK = Methyl ethyl ketone;
HCH = Cyclohexanone hydroperoxide;
HET = Hexachloroendomethylene tetrahydrophthalic acid.

(b) *Safety precautions when using epoxies.* Chemicals are potentially dangerous, so some precautions are necessary when handling them. Styrene can cause eye irritation and some accelerators and catalysts (peroxides) can cause skin irritation. Some fumes can be toxic and inflammable.

The following precautions should therefore be observed:

(i) Adequate ventilation of the prep room.
(ii) Use of protective goggles and gloves, if necessary.
(iii) No inflammable solvents to be used for cleaning.
(iv) No smoking and no naked flame.

25. Polyurethanes. The reaction between an isocyanate and an alcohol gives a urethane:

$$R.NCO + HO.R' \longrightarrow R.NH.COOR'$$

isocyanate   alcohol   urethane

With polyfunctional isocyanates and polyols (polyhydric alcohols), polyurethanes are formed. The reaction is not one of condensation but one of rearrangement polymerisation.

The polymers can be linear thermoplastics (with diisocyanates and dihydric alcohols) or cross-linked thermosets (with diisocyanates and trihydric alcohols).

Depending on the nature of starting materials and compound additions such as fillers, etc., polyurethanes can be produced in various forms:

Flexible foams (for upholstery, mattresses, textile linings, acoustical thermal insulation, crash padding in cars, packaging for delicate equipment, etc.).

Rigid foams (for thermal insulation, refrigeration, cold stores, building-panel cores, etc.).

Elastomers (for castings, joints and sealants, gaskets, shoe industry).

Liquids (for paints, coatings, clear finishes and adhesives).

The polyurethanes are characterised by high strength, high dimensional stability, high resistance to chemicals and solvents. However, they are expensive and affected by sunlight unless light-stabilised.

Cyanates are very toxic and the usual precautions (*see* **24**) must be observed.

Typical properties are given in Table LI.

	Rigid foam (sheets and blocks)	Flexible foam (sheets and blocks)
Density ($\times 10^3$kg m^{-3})	32–60	40–80
Softening point	150–185 °C	150–185 °C
Thermal conductivity (Wm^{-1}K^{-1})	0·020–0·025	0·035
Thermal expansivity ($\times 10^{-6}$ K^{-1})	20–70	50–70
Compressive strength (MN m^{-2})	0·17	0·02–0·08
Effect of sunlight (UV)	Affected, unless suitably stabilised	
Effect of chemicals	Resistant to chemicals; some are affected by alkali	
Water absorption (7 days) (% by vol.)	2·5	Up to 10 or higher
Sound absorption	Low	High
Processing	Chemical reaction between two liquid components	

SILICONES

This unique class of polymers is based not on carbon but on silicon. Silicones have a $-Si-O-Si-$ backbone and are one of the most important inorganic polymers. A variety of silicone polymers can be produced ranging from low viscous fluids (linear chain of relatively low molecular weight) to rigid cross-linked resins (three-dimensional network polymers of high molecular weight). They are characterised by high dimensional and high thermal stabilities, chemical inertness, and low toxicity, good electrical and anti-adhesive properties.

26. Silicone oils. Silicone oils are chain molecules with no cross-linking. They are produced by hydrolysis and condensation of an intermediate compound dimethyldichlorosilane:

$$n\ Cl\!-\!\underset{\underset{CH_3}{|}}{\overset{\overset{CH_3}{|}}{Si}}\!-\!Cl \xrightarrow[\substack{+\\ \text{condensation}}]{\text{hydrolysis}} \cdots\!-\!O\left(\!-\!\underset{\underset{CH_3}{|}}{\overset{\overset{CH_3}{|}}{Si}}\!-\!O\!-\!\right)_n\cdots$$

dimethyldichlorosilane   silicone oil

The viscosity of the oil increases with increase in n ($n = 2$–2000 approximately) and is stable up to 200 °C. The fact that silicone oils are inert, water-repellent and good lubricants under fluid film conditions, results in their use in hydraulic systems, as release agents, as ingredients in greases, polishes and paints and for waterproof treatment of papers and fabrics.

Some physical properties of silicone fluids are given in Table LII.

TABLE LII: PHYSICAL PROPERTIES OF SILICONE LIQUIDS

$$CH_3-\underset{\underset{CH_3}{|}}{\overset{\overset{CH_3}{|}}{Si}}-O\left(-\underset{\underset{CH_3}{|}}{\overset{\overset{CH_3}{|}}{Si}}-O-\right)\underset{\hat{n}CH_3}{\overset{\overset{CH_3}{|}}{Si}}-CH_3$$

Polymer $n \simeq$	23	90	210	350
Molecular weight	1900	6700	15 800	26 400
Viscosity (η) at 25 °C ($mm^2 s^{-1}$)	20	100	350	1000
Density: ($\times 10^3 kg\ m^{-3}$)	0·947	0·965	0·969	0·970
Refractive index	1·400	1·403	1·4032	1·4035

27. Silicone elastomers or rubbers. By increasing the value of n in the region 5000–10 000 a gummy material is obtained. By oxidation, cross-linking can take place to produce an elastomer or rubber:

In silicone rubbers, cross-linking takes place once for every 200 chain units.

Silicone rubbers have a wide service temperature range (-75 to $+250$ °C), outstanding resistance to weathering and chemicals, excellent electrical insulating properties and good anti-adhesive properties. They are highly resistant to ozone in the upper atmosphere.

Main uses include gaskets for refrigeration doors, aircraft applications (*e.g.* ducting for hot air, gaskets and sealing rings for jet engines), electrical insulation for cables and wires, encapsulation of electrical and electronic components, and even as adhesives and bonding agents.

Cellular silicone rubbers are available and are widely used in the aircraft industry as seals and gaskets.

28. Silicone resins. Silicone resins are obtained by hydrolysis and condensation of methyltrichlorosilane, followed by oxidation.

$$
\begin{array}{c}
CH_3 \\
| \\
Cl\text{—}Si\text{—}Cl \\
| \\
Cl
\end{array}
\quad \text{methyltrichlorosilane}
$$

hydrolysis $\big|$ + condensation

$$
\begin{array}{ccc}
| & | & | \\
-Si\text{—}O\text{—}Si\text{—}O\text{—}Si\text{—}O- \\
| & | & | \\
O & O & O \\
| & | & | \\
-Si\text{—}O\text{—}Si\text{—}O\text{—}Si\text{—}O- \\
| & | & |
\end{array}
\quad \text{(network polymer)}
$$

As a result of oxidation, cross-linking takes place and silicone resins result. The resins possess high service temperatures (up to 300 °C), chemical inertness, water-repellency and good electrical properties.

Main uses include release resins, water-repellent treatment for brickwork and concrete, laminating resins (for asbestos–paper laminates, silicone–glass laminates), encapsulation of electronic components, impregnation of electrical windings, coating of paper and fabrics.

Typical properties are given in Table LIII.

TABLE LIII: TYPICAL PROPERTIES OF SILICONES

	Silicone rubbers	Silicone resins
Density ($\times 10^3$kg m^{-3})	1·15–1·30	1·88
Heat-distortion temperature		Up to 450 °C
Maximum service temperature	250 °C	
Thermal conductivity (Wm^{-1}K^{-1})	0·17	0·50
Thermal expansivity ($\times 10^{-6}$K^{-1})		24–29
Tensile strength (MN m^{-2})	5–10	42
Compressive strength (MN m^{-2})		100
Elongation	30–80%	
UV resistance	Excellent	
Chemical resistance	Excellent	
Electrical insulation	Excellent	
Water absorption	0·09	
Processing	Transfer moulding, compression moulding	

RUBBERS

Rubbers are highly elastic and resilient polymeric materials which find wide applications in various fields.

29. Natural rubber. Natural rubber is obtained by acid treatment of the latex (exudation) of a type of tree (*Hevea braziliensis*) now widely cultivated in Malaysia. Chemically it is mainly composed of cis-polyisoprene (Fig. 35(*c*))—a linear high molecular weight polymer, having a coiled or spiral structure.

(*a*) *Additives.* Raw rubber can be compounded with other additives:

(*i*) *Fillers*—to reduce cost, improve bulk, strength and abrasion resistance, *e.g.* carbon black, china clay (reinforcing fillers); whiting, barytes, talc (non-reinforcing fillers).

(*ii*) *Plasticisers*—softening agents such as stearic or oleic acid, petroleum jelly, various resins.

(*iii*) *Pigments*—to give range of colours.

(*iv*) *Antioxidants*—to improve resistance to sunlight, heat, oils and solvents, *e.g.* some derivatives of amines and phenols.

Some types of amines can also act as antiozonants, which improve resistance to attack by ozone.

(b) *Vulcanisation of rubber.* Rubber can be cured at 100–200 °C by a cross-linking agent to form vulcanised rubber with improved mechanical strength and modulus of elasticity. Sulphur is the cross-linking agent most commonly used, although others (*e.g.* sulphur dioxide, benzoyl

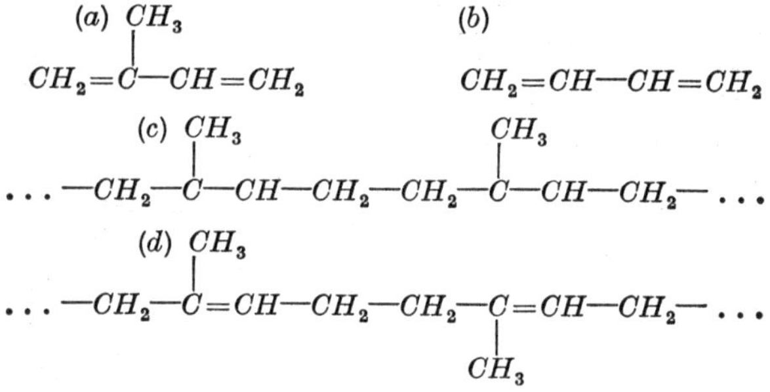

(a) CH_3
$CH_2{=}C{-}CH{=}CH_2$

(b)
$CH_2{=}CH{-}CH{=}CH_2$

(c) CH_3
$\ldots{-}CH_2{-}C{-}CH{-}CH_2{-}CH_2{-}C{-}CH{-}CH_2{-}\ldots$ $\quad CH_3$

(d) CH_3
$\ldots{-}CH_2{-}C{=}CH{-}CH_2{-}CH_2{-}C{=}CH{-}CH_2{-}\ldots$
$\qquad\qquad\qquad\qquad\qquad\quad CH_3$

FIG. 35.—*Natural rubber.*

(a) *Isoprene.*
(b) *Butadiene.*
(c) *Cis-polyisoprene (natural rubber).*
(d) *Trans-polyisoprene (gutta percha).*

peroxide, phenols, etc.) can be used. Cross-linking takes place between linear polyisoprene polymers at the unsaturated site, within the polymers (*see* Fig. 36). Depending on the amount of cross-linking agent added, the product may range from flexible rubbers (with 2–3 per cent sulphur) to hard rigid rubber called ebonite (with 30 per cent or more sulphur). The vulcanisation process is generally aided by accelerators (*e.g.* aniline, dithiocarbamates) or retarders (acetylsalicylic acid) and activators (zinc oxide and stearic acid). The properties also depend on the type of fillers and other additions used. Generally, natural rubbers have good resistance to alkalis and dilute acid, but are affected by most concentrated acids, aromatic and chlorinated hydrocarbons. They are used as solid rubber (car tyres, footwear), as ingredients in paint, coatings and adhesives and as electrical insulation for wires and cables.

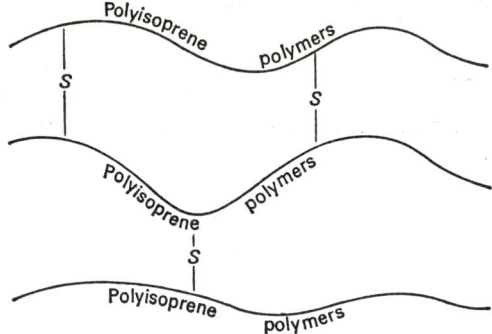

Fig. 36.—*Vulcanisation of rubber by sulphur.*

30. Synthetic rubbers. These are by no means substitutes for natural rubbers, but with improved technological know-how and techniques rubbers with specific properties can be designed and produced for particular needs and conditions.

(*a*) *Styrene–butadiene rubbers* (*SBR or Buna S rubbers*).

$$CH_2{=}CH{-}CH{=}CH_2 \qquad + \qquad CH_2{=}CH$$

butadiene

styrene

addition | polymerisation

$$-CH_2{-}CH{=}CH{-}CH_2{-}CH_2{-}CH{-}$$

SBR or Buna S rubbers

These are produced by copolymerisation of butadiene (*see* Fig. 35(*b*)) and styrene (commonly in the ratio 3:1).

They are very similar in properties to natural rubbers, except cheaper and more resistant to weathering but less resilient and resistant to tear. They are also used in car tyres, footwear, cable insulation, adhesives and moulded articles.

(*b*) *Isobutylene isoprene rubbers (IIR or Butyl rubbers).* These are made from isobutylene and isoprene (*see* Fig. 35(*a*)).

$$CH_2{=}\underset{\underset{CH_3}{|}}{\overset{\overset{CH_3}{|}}{C}} \quad + \quad CH_2{=}\overset{\overset{CH_3}{|}}{C}{-}C{=}CH_2$$

isobutylene   isoprene

addition | polymerisation

$$-CH_2{-}\underset{\underset{CH_3}{|}}{\overset{\overset{CH_3}{|}}{C}}{-}CH_2{-}C = \overset{\overset{CH_3}{|}}{C}{-}CH_2{-}$$

IIR or Butyl rubbers

They are also cheaper than natural rubbers but less resilient than SBR. However, they have a better resistance to sunlight, ozone, acids and oils, than SBR and natural rubbers. Their main asset is their extremely low permeability to gases.

They are used in tubeless tyres, mastics and sealants, electrical insulation of cables, etc.

(*c*) *Butadiene acrylonitrile rubbers (NBR, Buna N or Nitrile rubbers).* These are made from butadiene and acrylonitrile.

$$CH_2 = CH-CH = CH_2 \quad + \quad CH_2 = CH$$

butadiene

$$CN$$

acrylonitrile (15–40%)

additition | polymerisation

$$CH_2-CH = CH-CH_2 \left(-CH_2-CH- \atop CN \right)_n \cdots$$

Nitrile rubbers

Nitrile rubbers have good resistance to oils and organic solvents (better than natural and most synthetic rubbers except thiokol), better weathering resistance but poorer resilience and resistance to attack by ozone than natural rubbers. They are more expensive than SBR, butyl and natural rubbers.

Applications include gaskets, seals, tank linings, filler for phenol resins, ingredient for contact adhesives.

(d) *Polychloroprene rubbers (Neoprenes)*. These are made by polymerisation of chloroprene.

$$CH_2 = \overset{Cl}{\underset{|}{C}}-CH = CH_2 \xrightarrow[\text{polymerisation}]{\text{addition}}$$

chloroprene

$$\cdots(-CH_2-\overset{Cl}{\underset{|}{C}} = CH-CH_2-)_n$$

polychloroprene (Neoprene)

The chemical structure of neoprenes is similar to that of natural rubbers, with $(Cl-)$ replacing (CH_3-) group. Consequently resistance to fire is improved. Compared with natural rubbers, neoprenes are more resistant to atmospheric aging, sunlight, ozone and chemicals. They are more expensive than natural rubber, but cheaper than nitrile rubbers. Unlike the other rubbers mentioned, the vulcanising agent for polychloroprene is not sulphur but zinc oxide.

They are used in protective clothing, cable insulation, gaskets, adhesives, etc., but only to a limited extent in the tyre industry.

(e) *Polysulphides (Thiokol)*. These are made from ethylene dichloride and sodium polysulphide.

$$Cl\text{—}CH_2\text{—}CH_2\text{—}Cl + Na_2S_4 \xrightarrow[\text{polymerisation}]{\text{Condensation}}$$

$$\cdots \left(-CH_2\text{—}CH_2\text{—}\overset{\displaystyle S}{\underset{\displaystyle S}{\overset{\|}{\underset{\|}{S}}}}\text{—}S\text{—} \right)_n \cdots + 2\,NaCl$$

Thiokol *A*

Thiokols are extremely resistant to atmospheric weathering, ozone attack, oils, greases, and solvents, but their thermal stability at high or low temperatures is not as good as natural rubbers. They have an obnoxious odour which is characteristic of sulphides.

They are used in tank lining, sealing or caulking compounds, encapsulation of electrical components, moulded rubber articles, etc.

The properties of various types of rubbers are compared and summarised in Table LIV.

GLASS-REINFORCED PLASTICS (GRP)

Basically, resins in a more or less liquid state are reinforced with glass fibres either woven or in the form of mats and the process is carried out on a preformed mould or pattern followed by a period of curing (normally cold curing with thermoset resins). The simple fabrication process is that of "contact moulding" or "hand lay-up," which requires no expensive tools, equipment or plant.

31. Thermoset resins. The thermoset resins that can be used are:

(a) *Polyester resins*. Most widely used for GRP, commonly known under the trade mark of "fibreglass." The liquid resins and the catalyst systems are separately supplied.

TABLE LIV: COMPARATIVE PROPERTIES OF RUBBERS

Property	Type of rubbers							
	Natural	SBR	Butyl	Neo-prene	Nitrile	Thiokol	Silicone	Poly-urethane
Elasticity	α	α−	γ	β	β	γ	γ+	β
Atmospheric aging	β	β	α	α	α−	α+	α	α
Sunlight resistance	γ	γ	α−	α	β	α	α	α
Ozone resistance	β	β−	β+	α−	β−	α	α	α+
Flame resistance	γ	γ	γ	α	β	γ	α	γ
Oil resistance:								
Mineral	γ	γ	γ	β	β+	α+	γ	β+
Animal and vegetable	β−	β−	α	β+	α	α	β	α−
Acid resistance:								
Dilute	α	α−	α−	α+	α	α	α	β−
Concentrated	β−	β−	α	β+	β−	γ	β	γ
Water resistance	α	β+	α−	β	β	β	β	β+
Impermeability to gases	β	β	α+	β+	α−	α−	β	β+
Temperature resistance:								
Low	α	α−	α	β	β	α−	α+	α
High	β−	β	β+	β+	β+	γ	α+	β−

[*The Institute of Mechanical Engineers'*—*Handbook: Engineering Materials and Methods*

(b) *Epoxy resins.* The next in importance in the production of GRP. The resins are more expensive than the polyesters but are generally superior in many aspects, such as strength, dimensional stability and chemical resistance.

(c) *Phenolic resins.* Used in GRP particularly in situations where stiffness and strength at relatively high temperatures are essential. Curing generally requires heat and pressure.

(d) *Melamine formaldehyde resins.* Used to a limited extent in GRP. They are more costly but the composites possess outstanding dielectric strength and thermal stability at high temperatures. However, their mechanical strength is relatively weaker than the others.

(e) *Silicone resins.* Produce GRP which can withstand very high temperatures. High-temperature curing is required. They are expensive to produce.

(f) *Polyurethane resins.* Produce GRP which can have a high degree of flexibility.

32. Thermoplastic resins. It is also possible to use thermoplastic resins in developing glass-reinforced plastics. The type of thermoplastic resins which have been used satisfactorily are nylon, polypropylene, high-density polyethylene, polystyrene, polycarbonate and ABS.

33. General properties and applications of GRP. A very wide range of physical and chemical properties can be designed and produced from the wide choice of resins. Strength and stiffness are normally associated with brittleness, while flexibility and toughness are associated with low yield strength. In GRP it is possible to combine the high strength of glass fibres with the toughness of the resin matrix.

Compared with other plastics, glass-reinforced plastics generally have improved properties such as strength, chemical resistance, thermal resistance, fire resistance, electrical properties, etc.

One of the important characteristics of GRP is the fact that they can be easily moulded into complex shapes, often at ordinary atmospheric pressure and at room temperature. Coupled with their relatively high mechanical strength, they are widely used as structural engineering materials comparable, and sometimes superior in certain aspects, to other traditional materials such as timber, metals and concrete (*see* applications of GRP in Table XXXVII and **23**).

COMPOUNDING INGREDIENTS

Finished plastics materials are obtained by compounding the polymer (resin) with other ingredients such as fillers, plasticisers, stabilisers, dyes and pigments.

34. Fillers. Fillers are added to the resin:

  (*a*) to reduce the cost and/or extend the bulk of the product;
  (*b*) to enhance mechanical and other properties.

Some examples are:

  (*i*) *Fibrous type:*
  Asbestos.
  Glass fibre.
  Mica.
  Woodflour.
  Saw dust.
  Cork dust.

(*ii*) *Non-fibrous type:*
 Barytes.
 China clay.
 Carbon black.
 Talc.
 Zinc oxide.
 Calcium carbonate.
 Titanium dioxide.

35. Plasticisers. These are normally non-volatile solvents added to polymers to improve flexibility and processing properties.

Some examples are:

  (*i*) Phthalates (dibutyl phthalate for PVAc).
  (*ii*) Phosphates (tricresyl phosphate for PVC).
  (*iii*) Polyesters.
  (*iv*) Epoxies.
  (*v*) Nitrile rubbers.

36. Stabilisers. These are used to reduce the adverse effects due to heat, sunlight, ozone.

Some examples are:

  (*i*) White lead (for vinyl stabilisation).
  (*ii*) Barium–cadmium laurate (good heat stabilisers).
  (*iii*) Antioxidants such as β-naphthylamine, certain ketones and aldehydes.

37. Colouring materials. To give range of colours to finished plastics products.

Some examples are:

  (*i*) Inorganic pigments—zinc oxide, white lead, titanium dioxide.
  (*ii*) Organic pigments—phthalocyanines.
  (*iii*) Dyes.

FORMING PROCESSES

Thermoplastics are supplied in the fully polymerised state, whereas thermosetting plastics are supplied as two separate components—partly-polymerised resins and curing agents.

38. Compression moulding. Compression moulding is applicable to both thermosetting and thermoplastic resins. The method involves pressing the top half of the moulding plate on to the lower half which contains the moulding materials. Pressure and often heat may be maintained until the materials are cured.

Automatic equipment is, however, rather expensive.

Typical products include washing machine agitators, motor instrument panels, etc.

39. Transfer moulding. This technique is used for thermosetting resins mainly (MF, PF) but can be used for thermoplastic resins as well. It is a modification of the compression moulding technique. The moulding powder is preheated in an antechamber before being forced into a heated mould. Intricately-shaped articles can be moulded using this technique. Typical articles include motor car distributor caps, electrical plugs, etc.

40. Injection moulding. This technique is more suitable for thermoplastic than thermosetting resins. The moulding material is softened by heating in a cylinder and injected under pressure into a preformed mould where setting and hardening takes place. Typical products include domestic wares, crates and boxes, etc.

41. Blow moulding. This technique is used mainly for thermoplastics, usually polyethylene, PVC, polystyrene and polypropylene.

The principle is similar to that used in the production of glass bottles. It involves forcing air into a sealed molten body surrounded by a cold mould.

Typical products include bottles, containers, cold water cisterns, etc.

42. Extrusion. This is used mainly for thermoplastics (*e.g.* PVC, polyethylene, polystyrene) but can be used for thermosetting resins. The granular resins are fed through a hopper into a heated cylinder by the action of an Archimedean screw and through a die to give the required shape. It is a continuous process for producing pipes, rods, films or sheets, etc.

43. Thermoforming. This technique, mainly used for thermoplastics such as polystyrene, polyethylene and acrylics, involves softening a plastic sheet before being drawn into a female mould by use of vacuum or formed over a male mould by a vacuum (*vacuum moulding*). It is suitable for producing complicated shapes and profiles. Typical products are vending machine cups, refrigerator liners, aircraft canopies, cold water cisterns, plastic toys, etc.

44. Calendering. This is a fast and economical continuous method for producing films or sheets and is suitable for thermoplastics such as PVC. It is a modified form of extrusion without the use of a die but with the use of a system of rollers.

PROGRESS TEST 9

1. What is a polymer?
What are the main types of polymerisation reactions? (**1, 2**)
2. Show how the different polymerisation reactions can influence the properties of polymers in terms of structure and properties. (pp. 154–5)
3. Differentiate between thermoplastics and thermosetting resins. Give five examples of each type. (**3–4**)
4. Discuss the merits and limitations of plastics materials and compare them with those of other building or engineering materials. (**5–10**)
5. Outline the main applications of plastics in building. (Table **XXXVII**)
6. Compare the characteristic properties and uses of HD polythene and LD polythene. (**11**, Table **XXXVIII**)
7. What possible structures can be obtained from polypropylene? What effects have these structures on the mechanical properties of the product? (**12**, Table **XXXIX**)
8. What are the two types of PVC which are available commercially? Mention the advantage of copolymerising PVC. (**13**, Table **XL**)
9. What is the sole property that distinguishes between polystyrene and other plastics materials?
Name the various copolymers of polystyrene and state their main uses. (**14**)
10. What is PTFE?
What makes PVC and PTFE more fire-resistant than most other plastics materials? (**13, 15**)
11. Give the chemical name for "Perspex."
Mention some uses of perspex. (**16**)
12. Write down the chemical names for "Terylene" and "Nylons." (**17**)

13. What are polyacetals and polycarbonates?

Mention their characteristic properties and uses. (**18, 19**)

14. What are PF, UF and MF? Are they thermoplastics or thermosetting? (**21, 22**, Tables XLVII, XLVIII)

15. Polyesters can be thermoplastics or thermosetting, saturated or unsaturated. Illustrate and give at least one example of each. (**17, 23**)

16. What are epoxies? Show, by means of a schematic diagram, the general principle involved in the synthesis of epoxies and polyesters. (**24**)

17. Describe the various uses of polyurethanes. (**25**)

18. Distinguish between silicone oils, elastomers and resins in terms of (*a*) structure (*b*) properties. (**26–28**, Table LIII)

19. Explain the process of the vulcanisation of rubber. (**29**(*b*))

20. Write short notes on the following: (*a*) Styrene–butadiene rubbers, (*b*) Isobutylene isoprene rubbers, (*c*) Butadiene acrylonitrile rubbers, (*d*) Polychloroprene rubbers, (*e*) Polysulphides. (**30**)

21. List the thermoset and thermoplastic resins which can be used in the production of glass-reinforced plastics. (**31, 32**)

22. Give an account of the general properties and applications of glass-reinforced plastics. (**33**)

23. Mention the role of ingredients used in the production of the finished plastics materials. (**34–37**)

24. Give a brief account of the various forming processes commonly used for plastics materials. (**38–44**)

EXAMINATION QUESTIONS

1. Explain the principal difference between thermoplastic and thermosetting polymers.

What is meant by polymerisation of plastics?

Describe the processes involved in addition and condensation polymerisation.

What are the advantages of a copolymer?

(C.E.I. Part 2)

2. Differentiate between thermoplastics and thermosetting materials.

Give *two* common examples of each type and indicate their main characteristic properties.

Suggest *one* plastic material of your choice, with reasons, for use in the following situations:

    (*a*) Cold water storage cistern,

    (*b*) Hot-water pipes,

    (*c*) Shop display windows,

    (*d*) Dome roofs,

    (*e*) Bonding agents in plywood.

(P.S.B. Grad. Dip. Arch.)

3. (a) Distinguish, with examples wherever possible, between:

    (i) addition polymerisation and condensation polymerisation,

    (ii) linear, branching and cross-linked structures,

    (iii) random and alternating copolymers,

    (iv) isotactic, syndiotactic and atactic forms of polymers.

(b) Give an account of the main characteristics and uses of *either* rigid PVC *or* flexible PVC.

(S.O.E. Grad. Exam. C.E.)

4. (a) What is meant by "glass transition temperature" in polymer technology ?

(b) Describe briefly the effect on the glass transition temperature of the following:

    (i) branch substitution,

    (ii) steric hinderance,

    (iii) cross-linking,

    (iv) plasticisation.

Give one example in each case.

(c) Distinguish between the following stereostructures:

    (i) isotactic,

    (ii) syndiotactic,

    (iii) atactic.

Show, with the aid of examples, how they affect the properties of the polymers.

(P.S.B. B.Sc. S.E.)

5. Select, giving your reasons, *one* plastics material and *one* non-plastics material for use in the following situations:

    (i) Curing of concrete,

    (ii) Coldwater storage tank,

    (iii) Insulation of flat concrete roof,

    (iv) Rooflights,

    (v) Hotwater piping.

Discuss the disadvantages, if any, of the materials you selected.

(P.S.B. "A" Level Qualifying Course)

6. With reference to the physical and chemical properties of plastics, the requirements of occupancy, durability and cost, discuss the feasibility of producing an all-plastic house.

(P.S.B. B.Sc. Bldg.)

7. Briefly justify the use of FOUR plastics in current building practice with particular reference to the following factors:

    (a) weathering,

    (b) warmth,

    (c) water absorption,

(d) appearance,
(e) strength,
(f) fire resistance,
(g) workability,
(h) plasticity.

(I.O.B. Final Part 1)

8. Describe the formation of thermoset epoxide resins, and their uses (a) as modifiers for polysulphides and coal tar, (b) for surface treatment of roads and (c) in the repair of concrete.

(P.S.B. B.Sc. S.E.)

9. (a) Sketch the *four* types of flow curves commonly used to explain fluid behaviour, emphasising the nature of plastic flow and its difference from other flows.

(b) Which *three* fabricating techniques used in the metal industries have been successfully adapted for the manufacture of plastic articles?

(c) State (i) what is meant by a copolymer, (ii) the difference between thermoplastics and thermosets in terms of chemical bond and inter-molecular forces.

(d) Identify, in general terms, the nature and functions of six common ingredients in a commercial plastic mix for the manufacture of PVC floor-covering material.

(P.S.B. B.Sc. Bldg.)

10. (a) Describe *three* ways of establishing a polymer structure which has the characteristics of a rubber, and explain how it differs from an unvulcanised natural rubber.

(b) Describe rubber-modified polystyrene and rubber-modified asphalt, giving examples of their applications in civil and structural engineering.

(P.S.B. B.Sc. S.E.)

11. (a) Discuss the implications of the following components used in plastics technology:

(i) Filler,
(ii) Plasticiser,
(iii) Stabiliser,
(iv) Pigment,

and give an example of each component.

(b) Briefly explain and comment on the following forming methods for plastics:

(i) Extrusion moulding,
(ii) Vacuum forming.

(P.S.B. B.Sc. S.E.)

FURTHER READING

Evans, V., *Plastics as Corrosion-resistant Materials*, Pergamon, 1966.

Palin, G. R., *Plastics for Engineers*, Pergamon, 1967.

Freeman, G. G., *Silicones*, Plastics Institute, 1962.

Parkyn, B., *Glass Reinforced Plastics*, Iliffe Books, 1970.

Wilson, C., *Design Engineering Handbook—Plastics*, Product Journals Ltd., 1968.

E.E.U.A. Handbook No. 31: 1973—*The use of Plastics Materials in Building*, Constable.

CHAPTER X

SURFACE COATINGS

INTRODUCTION

This chapter will be mainly concerned with protective and decorative surface coatings for such materials as metals, timber, concrete, masonry, etc. Coatings include paints, varnishes and lacquers.

Paint—a pigmented material, which when applied in a liquid form to a surface, forms after a time a dry adherent film.

Varnish—a transparent coating composition based essentially on drying oils, resins and solvents.

Lacquer—a coating composition which dries solely by evaporation of the solvents, *e.g.* cellulose and spirit lacquers.

COMPOSITION OF PAINTS

The composition of paints and varnishes is shown schematically in Fig. 37.

1. Base or base pigment. A base pigment is the dispersed finely-powdered solid material in a paint, which imparts colour and opacity to the dried film. It includes white or coloured pigments (stainers) and extenders or fillers.

(a) *Pigment or stainer.* This is the solid powder (organic or inorganic in nature) which imparts the following characteristics to the paint: colour, opacity, durability (*e.g.* resistance to abrasion), chemical resistance (*e.g.* resistance to corrosion) and improved body.

Table LV gives various types of pigments.

(b) *Extender or filler.* A finely-powdered material (inorganic) which has a low refractive index and consequently little obliterating power, but is used as a constituent of paints to adjust certain properties, notably working and

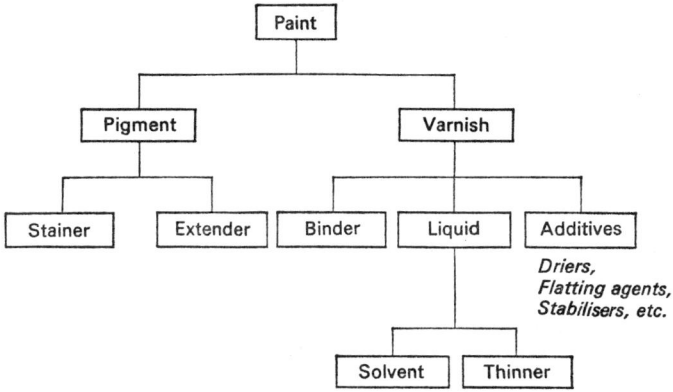

FIG. 37—*Composition of paints and varnishes.*

film-forming properties, and to avoid settlement on storage (BS 2015).

It is normally white in colour and there is no organic extender. Some extenders used are:

Barytes (white or off-white) $BaSO_4$	Chemically inert
Blanc fixe (precipitated $BaSO_4$)	Improves working properties
Whiting (Paris white) $CaCO_3$	Attacked by acids
Gypsum ($CaSO_4 . 2H_2O$)	Usually in combination with titanium dioxide
China clay (kaolin— $Al_2O_3 . 2SiO_2 . 2H_2O$)	Decomposed by hot concentrated acids
Silica dust (SiO_2—Quartz, Flint, Kieselguhr)	Cheap
Slate powder (impure hydrated aluminium silicate)	Dirty colour
Mica ($K_2O . 2Al_2O_3 . 6SiO_2 . 2H_2O$)	Increases brilliance of coloured pigments
Bentonite (hydrated aluminium silicate—$Al_2O_3 . 4SiO_2 . 2H_2O$, Al replaceable by Mg and Na)	Used also as thickeners in water paints

2. Vehicle or medium (varnish)

(a) *Binder or film-former*. The non-volatile portion of the vehicle of a paint; it binds or cements the pigment particles

TABLE LV: PIGMENTS

Colour	Types	Characteristics
White	Titanium dioxide TiO_2 (Anatase type and Rutile type)	Very high refractive index (n), chemically inert and physiologically non-toxic Rutile type is very resistant to chalking $SG = 3\cdot8-4\cdot2$; $n = 2\cdot52-2\cdot76$
	White lead $Pb(OH)_2.PbCO_3$	Good compatibility with oil, very elastic and highly resistant to moisture Chalks slowly and in polluted atmospheres gradually turns to black sulphide Toxic $SG = 6\cdot7$; $n = 2\cdot0$
	Zinc oxide ZnO	Relatively reactive, causing cracking with oleoresinous finishes Absorbs UV radiation Fungicidal but non-toxic to man $SG = 5\cdot65$; $n = 2\cdot0$
	Lithopone $ZnS + BaSO_4$	Excellent hiding power Liable to chalking and weathering by erosion $SG = 4\cdot2$; $n = 1\cdot9-2\cdot2$
	Antimony oxide Sb_2O_3	Improves brushing and levelling properties $SG = 5\cdot5-5\cdot7$; $n = 2\cdot0-2\cdot2$
	Calcium plumbate $2CaO.PbO_2$	Light buff colour Suitable as primer for ferrous metals and wood
Blue and Green	Prussian Blue $Fe_4(Fe(CN)_6)_3$	Reaction with alkalis $SG = 1\cdot97$
	Ultramarine blue—complex silicate (a natural material)	Attacked by acids, but resistant to alkalis Widely used in distempers and water paints
	Cobalt blue $CoO.Al_2O_3$	Clear and almost transparent colour Resistant to heat, light, acids and alkalis
	Brunswick greens or Chrome greens Cr_2O_3	$SG = 5\cdot2$
Brown and Red	Iron oxides: Siennas (hydrated Fe_2O_3 + clay) Umbers ($Fe_2O_3 + MnO_2$) Synthetic Fe_2O_3	} Natural earths e.g. Turkey red, Indian red
	Cadmium reds $CdS.CdSe$	Resistant to alkalis, organic acids and dilute mineral acids Expensive Non-toxic $SG = 4\cdot5-4\cdot7$
	Chrome reds (lead chromate + lead sulphate + lead molybdate)	Toxic Gradually converted to sulphide in polluted atmospheres
	Red lead Pb_3O_4	Primers for ferrous metals $SG = 8\cdot8$; $n = 2\cdot42$
Orange and Yellow	Iron oxides: Yellow ochres (hydrated Fe_2O_3 + silicates)	Natural earths
	Lead chromates: (based on $PbCrO_4$)	All toxic Colour shades dependent on $PbCrO_4$ content
	Barium chromate $BaCrO_4$	Soluble in dilute hydrochloric acid Used as stopper on light alloys Resistant to alkalis
	Cadmium sulphide CdS	Insoluble in paint medium (similar to Cadmium reds) $SG = 4\cdot2-4\cdot3$

Colour	Types	Characteristics
	Zinc chrome $4ZnCrO_4.K_2O.H_2O$	Poor opacity Suitable as primers for light alloys and steel $SG = 3.4-3.5$
	Titanium yellow	
Black	Black iron oxide Fe_3O_4 Carbon black	Chemically inert Light but hard to grind Excellent staining power $SG = 1.8$
Organic pigments	(No organic white or extender) Toluidine reds Toluidine yellow Phthalocyanine greens and yellows Aniline black	
Metallic pigments	Aluminium Al (powder or flakes)	High-temperature resistance Non-toxic Suitable as primers for metals and wood
	Zinc Zn (flakes or dust)	Good opacity, but very reactive Suitable as primer for iron and steel structures (acting as sacrificial anode) Non-toxic
	Lead Pb	Toxic
Luminous pigments	(i) *Fluorescent pigments:* Calcium or Magnesium Tungstates Zinc orthosilicates Cadmium borates (ii) *Phosphorescent pigments:* Zinc Sulphide ZnS Calcium sulphide CaS Strontium Sulphide SrS Barium Sulphide BaS Cadmium Sulphide CdS	Yellowish green Violet Blue/green Yellow Red

together, and the paint film as a whole to the material to which it is applied (BS 2015). The drying of the binder into a hard glossy film takes place either by evaporation or by polymerisation/oxidation reaction. Some examples are given in Table LVI.

(b) *Driers.* These are substances which, when incorporated in relatively small proportions in drying oils, or in paints or varnishes based on drying oils, bring about an appreciable reduction of drying times at ordinary temperatures (BS 2015). Driers are catalysts added to accelerate the drying process. The requirements of a dryer are consistency in quality and colour, stability and compatibility.

Examples are the naphthenates and linoleates of cobalt, manganese, calcium and lead.

TABLE LVI: BINDER OR FILM-FORMER

Drying by	
Evaporation	Polymerisation
Shellac	Drying oils (linseed, tung,
Gum	soya bean oils)
Bitumen	Alkyd resins
Cellulose acetate	Phenolics (PF)
(Thermoplastic) alkyd resins	Aminoplasts (UF, MF)
Chlorinated rubber	Epoxy resins
Vinyl resins	Polyurethanes
	Polyester resins
(Thermoplastic) acrylic resins	Silicone resins

(c) *Thinners and solvents.*

(i) *Thinners* are volatile liquids added to paints and varnishes to facilitate application and to aid penetration by lowering viscosity. They should be completely miscible with the paint or varnish at ordinary temperatures and should not cause precipitation of the non-volatile portion either in the container or in the film during drying (BS 2015).

Examples are: turpentine, white spirit, naphtha, water.

(ii) *Solvents* also are volatile liquids added to paints in order to dissolve or disperse the film-forming constituents. They evaporate during drying and are used to control the consistency and character of the finish and to regulate application properties.

Examples are alcohols, esters, ethers, ketones, chlorinated hydrocarbons.

(d) *Additives.* Some other additives are usually used.

(i) *Flatting agents* are materials added to paints to reduce the gloss of the dried film.

Some examples include aluminium stearate, fine silica particles, silicates.

(ii) *Anti-settling agents* are added to stabilise dispersion and delay settlement on storage. They include stearates and palmitates of aluminium, zinc and calcium.

(iii) *Anti-skinning agents* are added to prevent or retard the polymerisation and/or oxidation processes which result in the formation of an insoluble skin on the surface of the paint in a container.

An example is methylethyl ketoxime.

PAINT COAT SYSTEMS

A paint system is a multiple coat system in which each coat fulfils a specific function. A simple paint system involves (*a*) primary coat or primer, (*b*) undercoat and (*c*) finishing coat.

3. Primary coat or primer. This is the first complete coat of paint of a painting system applied to an unpainted surface. The type of primer varies with the surface, its condition and the painting system to be used (BS 2015).

The main purpose of a primer is to ensure maximum and lasting adhesion to the surface of objects (substrates) and to subsequent coats. In addition, a primer provides protection of the substrate against corrosion (in the case of metals) or against moisture and weathering (in the case of porous materials such as timber and timber products).

Table LVII gives some examples.

Some typical formulations of primers are given in Table LVIII.

TABLE LVII: PRIMERS (CP 231)

	Medium or binder	*Suitable for*
Primers for metals		
Red lead	Linseed oil (BS 2523) Quicker-drying media, *e.g.* alkyd, epoxy ester or oleoresinous, may be used	Iron and steel
Calcium plumbate	Linseed oil (BS 3698)	Iron and steel, galvanised iron, zinc (wood)
Zinc chrome	Oil, oleoresinous or alkyd media	Aluminium, zinc, iron and steel
Zinc-rich	Polystyrene, chlorinated rubber, polyamide-cured epoxy ester, polyurethane or styrenated alkyd	Iron and steel
Metallic lead	Linseed oil, chlorinated rubber or oleoresinous media	Iron and steel
Primers for wood		
Pink	Red and white lead pigments in linseed oil (BS 2521), oleoresinous or alkyd media	Wood (softwood)
Leadless (TiO_2)	Oil, oleoresinous or alkyd media	Wood (softwood)
Aluminium	Oleoresinous and/or alkyd media	Wood (softwood and hardwood), metals (copper)
Primers for plasters or other wall surfaces		
Alkali-resistant	Tung oil/phenolic resins, tung oil/coumarone resins, or chlorinated rubbers (pigment: anatase TiO_2)	Plasters, brickwork, concrete, asbestos board
Sharp	White lead pigment in gold size or boiled linseed oil	Plasters, Keene's cement, asbestos board

TABLE LVIII: TYPICAL FORMULATIONS OF PRIMERS

	Percentage by weight	Function
Zinc-rich primer—polystyrene base		
"Superfine" zinc dust	75·10	Pigment
Vinalak 5740 (polystyrene resin)	2·96	Binders
Arochlor 1254 (chlorinated diphenyl)	0·99	Plasticiser/ flame retardant
Bentone 34	0·39	Thickener
Xylol	11·86 ⎫	Solvents
Solvent naphtha	8·70 ⎭	
	100·00	
Wood primer		
White lead in oil (6–11% oil)	50·00 ⎫	Pigment
Titanium dioxide	10·00 ⎭	
China clay	12·00	Extender
Calcium carbonate (surface coated with stearate)	3·00	Extender/anti-settling additive
Linseed oil–rosin-modified phenolic resin at 50% solids (70% oil length)	10·00 ⎫	Binders
Boiled linseed oil	5·00 ⎭	
Lead naphthenate (24% Pb)	0·20 ⎫	Driers
Cobalt naphthenate (1% Co)	1·00 ⎭	
White spirit	8·80	Solvent
	100·00	
Primer for brickwork (ICI Mond 11D 182/1)		
Alloprene R20	19·5	Binder
Cereclor 54	13·0	Plasticiser
Tioxide R–CR2	16·5	Pigment
Thixatrol ST	0·5	Anti-settling agent
Aromasol H	40·4 ⎫	Solvent
White Spirit	10·1 ⎭	
	100·0	

4. Undercoats. An undercoat is the coat or coats applied to a surface after priming, filling, etc. or after the preparation of a previously-painted surface, and before the application of a finishing coat. An undercoat should possess good hiding power and a colour leading up to that of the finishing coat, and should be suitable for use with other paints in the system (BS 2015).

Undercoats must be compatible with the priming and finishing coats used. They are usually low-gloss and highly-pigmented.

Undercoats may be based on long oil/phenolic media or long oil alkyds (for building undercoats), on alkyd or epoxy blends with urea resins (for automobile undercoats), or on acrylic resins as primer/undercoats suitable for industrialised buildings.

A typical formulation of a building undercoat is given in Table LIX.

TABLE LIX: TYPICAL FORMULATION OF A BUILDING
UNDERCOAT

Air-drying building undercoat, white (Cray Valley Products Ltd. formulation 1339 A)	Percentage by weight	Function
Rutile titanium dioxide	23·80	Pigment
Barytes	14·00 ⎫	
China clay	5·10 ⎬	Extenders
Micronised dolomite	10·60 ⎭	
Synolac 44 WHF (54% soya rosin-modified alkyd)	25·30 ⎫	Binders
Gelkyd 310 W (Thixotropic 47% soya rosin-modified alkyd)	9·40 ⎭	
Methyl ethyl ketoxime	0·06	Anti-skinning agent
Lead naphthenate (24% Pb)	0·35 ⎫	Driers
Cobalt naphthenate (6% Co)	0·14 ⎭	
White spirit	3·00 ⎫	Solvents/thinners
Petroleum distillate	8·25 ⎭	
	100·00	

5. Finishing coats (or finishes). A finishing coat is the final coat in a paint system. Its purpose is to provide colour, gloss or texture to the painted surface, as well as protection from weathering. The degrees of glossiness can be described as high gloss, semi-gloss, eggshell or matt finish.

High-gloss finishes are based on oil, oleoresinous or alkyd media or on blends of these.

Semi-gloss and flat finishes are for internal use only and are based on oil, oleoresinous or alkyd media with flatting agents.

TABLE LX: TYPICAL FORMULATIONS OF FINISHES

	Percentage	*Function*
Alkyd gloss finish		
Titanium dioxide	22·00	Pigment
Blanc fixe ($BaSO_4$)	5·00	Extender
Soya bean oil/alkyd in white spirit	60·00	Binder
Lead naphthenate (24% Pb)	2·00 ⎫	Driers
Cobalt naphthenate (1% Co)	2·00 ⎬	
White spirit	9·00 ⎭	Solvent
	100·00	
Inexpensive oleoresinous gloss finish		
Titanium dioxide	12·00 ⎫	
Yellow oxide	0·50 ⎬	Pigments
Midchrome yellow	0·20 ⎭	
Whiting	23·00	Extender
Linseed oil, rosin-modified	45·00 ⎫	
Phenolic resin in white spirit		Binders
Bodied linseed oil (of viscosity		
4NSm^{-2})	3·00 ⎭	
Lead naphthenate (24% Pb)	0·80 ⎫	Driers
Cobalt naphthenate (1% Co)	0·50 ⎭	
White spirit	15·00	Solvent
	100·00	

Other polymeric media (including bituminous types) can be used.

Typical formulations are given in Table LX.

6. Special types of paints.

(*a*) *Industrial maintenance paints.* These are developed to resist severe atmospheric corrosion and weathering. They are normally based on chemically inert pigments (*e.g.* zinc, titanium dioxide, red oxide) dispersed in chlorinated rubber or cross-linked epoxy resins.

TABLE LXI: TYPICAL FORMULATIONS OF INDUSTRIAL
MAINTENANCE PAINTS

	Percentage	*Function*
Zinc dust brush primer (ICI Mond 11A1)		
Zinc dust	74·80	Pigment
Alloprene R90 (chlorinated rubber)	3·90 ⎫	Binders
Cereclor 48 (chlorinated paraffin wax)	2·60 ⎭	
Modified hydrogenated castor oil	0·20 ⎫	Anti-settling agents
Epoxidised soya bean	1·00 ⎭	
Aromasol H (high-boiling paint solvent)	17·50	Solvent
	100·00	
Inert high build paints for brush application (ICI Mond 11G 216/1)		
Alloprene R10	19·6	Binder
Cereclor 70	13·0 ⎫	Plasticisers
Cereclor 42	6·7 ⎭	
Thixatrol ST	1·7	Anti-settling agent
Tioxide R–CR2	13·3 ⎫	Pigments
Carbon black–tinting	0·2 ⎭	
Barytes	13·3	Extender
Bentone 34	0·1	Thickener
Xylene	16·0 ⎫	Solvents
Aromasol H	16·1 ⎭	
	100·0	

(b) *Fungus-resisting paints.* To control and suppress fungal growths, *e.g.* mould and mildew in humid situations, fungus-resisting paints are used. They contain one or more fungicidal additives in their formulation. Copper and mercury compounds are common fungicides or preservatives used.

A typical formulation is given in Table LXII.

TABLE LXII: TYPICAL FORMULATION OF A
FUNGICIDAL GLOSS PAINT

White fungicidal gloss paints	Percentage	Function
Tioxide R-CR	26·90	Pigment
Zinc oxide	4·00	Pigment/ fungicide
70% linseed alkyd vehicle	54·20	Binder
Lead naphthenate (24% Pb)	3·00	Driers
Cobalt naphthenate (6% Co)	0·30	
Nuodex 321 extra (Mercurial fungicide)	0·20	Fungicide (toxic)
White spirit	11·40	Solvent
	100·00	

(c) *Anti-fouling paints*. To prevent marine growths, *e.g.* on ships' hulls, anti-fouling paints have been developed. They contain toxic additives such as mercury and copper compounds.

A typical formulation is given in Table LXIII.

TABLE LXIII: TYPICAL FORMULATION OF AN
ANTI-FOULING PAINT

Anti-fouling composition	Percentage	Function
Cuprous oxide	44·30	Pigment
Iron oxide	4·70	
Vinylite VAGH (Vinyl chloride copolymer)	12·00	Binder
Tricresyl phosphate	3·80	Plasticiser
Methyl isobutyl ketone	21·10	Solvents
Xylol	14·10	
	100·00	

(d) *Thixotropic paints*. They have jelly-like consistency and non-drip properties. They are free-flowing when brushed and are less likely to run or sag. Consistency is reduced by stirring or brushing out, but reverts to original state on standing.

A typical formulation is given in Table LXIV.

TABLE LXIV: TYPICAL FORMULATION OF A
THIXOTROPIC PAINT

Brilliant white jelly gloss paint (Cray Valley Products Ltd. formulation)	Percentage	Function
High opacity rutile titanium dioxide	35·70	Pigment
Gelkyd 357W (polyamide-treated alkyd)	58·40	Binder
Lead naphthenate (24% *Pb*)	0·73	Driers
Cobalt naphthenate (6% *Co*)	0·17	
Methyl ethyl ketoxime	0·17	Anti-skinning agent
White spirit	4·83	Solvent
	100·00	

(e) *Heat-resisting paints.* Paints when subjected to elevated temperatures over a prolonged or continuous period are liable to discolour and/or become brittle. Conventional decorative paints based on alkyd or oleoresinous media can be safely used on surface temperatures below the boiling point of water (*e.g.* on central heating radiators). For higher temperatures, heat-resisting paints based on media such as silicone alkyd resins should be used.

(f) *Fire-retardant paints.* These are suitable for application on timber, hardboard, chipboard, etc. which are combustible. They are usually based on chlorinated rubber or synthetic resin emulsions containing ingredients which have fireproofing characteristics.

7. Emulsion paints. These are paints in which the binder (a synthetic resin) is dispersed as fine droplets in water. Other important ingredients include pigments, dispersing or wetting agents and stabilisers.

The binders in common use are *PVAc*, *PVAc* copolymer, acrylic copolymer and styrene–butadiene copolymer emulsions.

A typical formulation is given in Table LXV.

TABLE LXV: TYPICAL FORMULATION OF AN
EMULSION PAINT

Emulsion paint based on PVAc	Percentage	Function
Titanium dioxide	22·30	Pigment
China clay	10·00	Extender
Whiting	1·50	Alkalinity control
Sodium hexametaphosphate	0·20	Dispersion agent
Polyoxyethylene ether of octylphenol	0·50	Surfactant
Sodium carboxymethyl cellulose	0·30	Thickener
Phenyl mercury acetate (10% Hg)	0·20	Fungicide
Water	30·00	Diluent
PVAc water plasticised with 15% dibutyl phthalate	35·00	Binder
	100·00	

PAINT APPLICATION

8. Surface preparation and pretreatment. Surface preparation prior to painting is a very important step towards achieving full decorative and protective value from the process of painting.

Its purpose is to obtain a clean surface free from grease, loose particles or corrosion products (especially in the case of metals), to even out any irregularities in the surface and, in the case of smooth non-adherent metallic surfaces, to ensure mechanical adhesion of the paint by chemical treatment.

(a) *Iron and steel.* Surface preparation prior to priming involves:

   (i) Cleaning by use of solvent (*e.g.* white spirit).
   (ii) Removal of mill scale and rust (*e.g.* by wire brushing, acid pickling, sand blasting or by application of oxyacetylene flame).
   (iii) Pretreatment with phosphating solutions (metal phosphates in phosphoric acid) to provide a passive surface.

(b) *Non-ferrous metals.*

   Aluminium —clean with organic solvent and pretreat with
   and alloys phosphate or chromate solutions.

Zinc and alloys	—clean with organic solvent and pretreat with phosphate or chromate solutions to remove basic zinc carbonate surface film.
Magnesium and alloys	—clean with organic solvent and pretreat with chromating solutions or anodise in a solution of ammonium bifluoride containing a small amount of chromate.
Cadmium and alloys	—clean and pretreat (as for zinc).
Copper and alloys	—abrade with glasspaper, clean with organic solvent and pretreat with etch primer (based on alcoholic phosphoric acid and basic zinc chromate pigment).
Lead and alloys	—abrade, clean and apply etch primer (as for copper).

(c) *Timber and timber products.*

| Wood and plywood | —make sure that wood is properly seasoned. Rub down to a smooth finish, dust off, wash with solvent, treat knots, seal and prime. |
| Building boards | —clean and make sure the surface is dry prior to priming. |

(d) *Plaster.* Allow new surfaces to dry, remove any efflorescence by brushing, fill cracks, smooth surface, dust off and clean. Use alkali-resistant primer or emulsion paint.

(e) *Concrete.* Allow surfaces to dry, wire brush to remove loose particles, wash with water and again allow to dry. Use alkali-resistant primer or emulsion paint.

(f) *Brickwork and masonry.* Allow surfaces to dry, remove any efflorescence (especially for new surfaces) by brushing, dust off and clean. Apply alkali-resistant primer or emulsion paint.

9. Methods of paint application. The choice of methods of paint application depends on various factors, such as type of finish, type of components or surfaces to be painted. The methods mostly used are brushing, spraying and dipping.

(a) *Brush and roller application.* Brushes are made from natural bristles or synthetic (nylon) filaments. Brushing is a relatively slow process requiring some patience and craftmanship and can be used for most types of paints, particularly the slow-drying type. The main defects are poor levelling (brush marks) and sagging.

The use of rollers (made from lambs' wool or soft plastic

foam) and trays is another method of hand application. This technique is faster and requires less skill.

(b) *Spray application.* This technique involves atomising the paint into fine sprays (*e.g.* by compressed air) which are discharged from a spray gun on to the surface to be painted. The process is continuous, fast and suitable for complex objects. However, the main defects which occur are the "orange peel" appearance of the film and sagging. Various types of spray application are available:

(i) *Conventional air spray.*

(ii) *Airless spray* (no air used, pressure applied directly to paint).

(iii) *Hot spray* (use of temperature 60–70 °C).

(iv) *Electrostatic spray.* Here the paint spray is charged electrostatically (high voltage required) and deposited on the articles to be painted. It is not suitable for coating the inside of a hollow article, because of the fact that no charge can exist inside a hollow conductor.

(c) *Dip application.* The method involves dipping articles in a dip tank for a period before taking them out and allowing to drain. It is very simple and economical.

Various processes are:

(i) *Hand dipping.*

(ii) *Automatic dipping* (*or conveyorised dipping*). The *Roto-Dip* process is used for the application of primers to car bodies.

Main defects are fat edge and tears.

(iii) *Flow coating.* The articles are carried by a conveyorised system through an enclosed chamber containing a number of strategically-placed jets of paint. Smaller amounts of paint are required but greater solvent losses are involved.

Suitable for paints based on styrenated alkyd or medium-/ short-drying alkyds.

(iv) *Curtain coating.* This is an improved form of flow-coating. It is suitable for large objects and has the advantage of speed of application and saving in labour and materials.

(d) *Electrodeposition* (*or electrocoating*). The principle is one of electrolysis. The article to be painted is made the anode, the metal bath is the cathode and the electrolyte is the paint based on water-soluble resin (*e.g.* epoxy esters or alkyd-modified phenolics). A D.C. voltage of 40–250 volts is applied. The large anions of the resin migrate and form a coating on the anode.

The operation of the process is easy to control and a complete coverage of exposed surfaces (including sharp edges) with a uniform film can be obtained.

Car bodies are coated by this method.

PAINT FAILURES

Paint failures or defects are mainly due to:

    (*i*) Inadequate preparation of the surface.
    (*ii*) Choice of unsuitable paint or paint system.
    (*iii*) Bad workmanship.

Typical paint failures are summarised below—all definitions come from BS 2015.

10. Blistering.

(*a*) *Definitions.* Formation of dome-shaped projections or blisters in paints or varnish films by local loss of adhesion and lifting of the film from the underlying surface. Such blisters may contain liquid, vapour, gas or crystals.

(*b*) *Causes.* Excessive moisture or resin exudation (wood); rust (iron and steel).

(*c*) *Remedy.*

    (*i*) *Wood.* Allow to dry sufficiently, remove knots and resins, and clean prior to priming.

    (*ii*) *Iron and steel.* Clean, remove rust, pretreat and prime with anti-corrosive primer.

11. Brush marks (or ropey finish).

(*a*) *Definition.* Ridges in a dried paint film left by the brush.

(*b*) *Causes.* Poor levelling properties of paint or varnish; overbrushing.

(*c*) *Remedy.* Complete reformulation of paint.

12. Chalking.

(*a*) *Definition.* The formation of a friable, powdery coating in a surface of a paint film caused by disintegration of the binding medium.

(*b*) *Causes.* Effect of ultra-violet radiation (sunlight).

(*c*) *Remedy.* Abrade, clean and repaint.

13. Chipping.

(*a*) *Definitions.*

> (*i*) Removal of paint or rust and scale by mechanical means.
> (*ii*) The total or partial removal of a dried paint film in flakes by accidental damage.

(*b*) *Causes.* Poor adhesion, inferior nature or incompatibility of primer.

(*c*) *Remedy.* Strip, wash, dry and repaint.

14. Cracking.

(*a*) *Definition.* Splitting of a dry paint or varnish film, usually as a result of aging.

Various types of cracking are described as haircracking, checking, crazing, crocodiling.

(*b*) *Causes.* Failure of a coat or a system of coats (aging and loss of extensibility); application of hard-drying gloss over soft undercoat.

(*c*) *Remedy.* Strip, wash, dry and repaint.

15. Efflorescence.

(*a*) *Definition.* Development of crystalline deposit on surface of brick, cement, etc. due to water, containing soluble salts coming to the surface and evaporating so that the salts are deposited. In some cases, the deposit may be formed on the top of any paint film present, but usually the paint film is pushed up and broken by the efflorescence under the coat.

(*b*) *Causes.* Soluble salts.

(*c*) *Remedy.* Strip, wire brush, clean, allow to dry and repaint using alkali-resisting primer or emulsion paint.

16. Flaking.

(*a*) *Definition.* Lifting of the paint film from the underlying surface in the form of flakes or scales.

(*b*) *Causes.*

> (*i*) Moisture and/or loss of adhesion (wood).
> (*ii*) Detached scale or rust (iron and steel).
> (*iii*) Smoothness of surfaces (galvanised steel).

(*iv*) Softness of metal (lead).

(*v*) Efflorescence (brickwork, masonry).

(*c*) *Remedy*. Stop the cause of dampness and condensation. Strip loose paint, clean, wash, allow to dry and repaint.

17. Wrinkling or rivelling.

(*a*) *Definition*. Development of wrinkles in a film during drying, usually due to the initial formation of a surface skin.

(*b*) *Causes*. Expansion of paint film.

(*c*) *Remedy*. Abrade, smooth surface and repaint.

PAINT TESTING

Testing and evaluation of paints are carried out on liquid paint and on dried film.

18. Liquid paint.

(*a*) *Viscosity: Coefficient of viscosity* (η) *or simply viscosity*. This is a measure of the flow properties of a fluid and is defined as:

$$\eta = \frac{\text{Shearing stress}}{\text{Rate of shear or velocity gradient}}$$

Depending on the principle of measurement, two types can be distinguished—

(*i*) dynamic viscosity, which is based on Poiseuille's formula (capillary flow method) and (*ii*) kinematic viscosity which is based on Stokes' formula (falling sphere method) (BS 188).

(*i*) *Tube viscometers* and *rising bubble viscometers*, based on (*i*), are suitable for Newtonian paints and varnishes, but not for thixotropic paints.

(*ii*) *Falling sphere viscometers* and *Höppler falling sphere viscometers*, based on (*ii*), are suitable for Newtonian fluids (homogeneous).

(*iii*) *Efflux viscometers* (flow cup method—BS 3900: Part A6). The time taken by a given volume of liquid to flow through an orifice is a measure of the viscosity of the liquid. This method is suitable for quality control and can be used for industrial finishes and paints with good flow properties. It is not suitable for thixotropic paints.

(*iv*) *Rotational viscometers* are suitable for both Newtonian and non-Newtonian fluids and can be used for thixotropic paints.

(*b*) *Colour* (BS 3900). Colour of varnishes and clear liquids can be measured by spectro-photometers or colorimeters which incorporate the use of a photocell to measure reflectance or transmittance as a function of wavelength.

(*c*) *Fastness to light* (BSS 1215, 2661–86, 1006, 950). Paint-coated panels are exposed to a carbon arc in a rotating drum. At regular intervals, the exposed and unexposed surfaces are compared for change in colour. The Fugitometer is one instrument used for this test.

(*d*) *Opacity* (*or hiding power*). A simple method is to determine the amount of paint required to obscure uniformly the pattern on a card printed with black and white squares.

More sophisticated techniques have been developed, such as the Pfund cryptometer.

(*e*) *Drying time*. A simple method is to touch the film gently with the finger at intervals of time until no sticking occurs.

Other methods have been developed, such as dropping graded silver sand on to the surface, leaving for one minute before lightly brushing off. The drying time is taken at the point when no damage is done on brushing the sand off the surface.

19. Dried film.

(*a*) *Flexibility and extensibility*. Flexibility is defined as the degree to which a paint film, after drying, is able to conform to movement or deformation of its supporting surface, without cracking or flaking (BS 2015).

Flexibility is related more to extensibility of the film than adhesion. A measure of flexibility of the film is determined by the *Mandrel Tests* (Bend Tests), which involve bending a painted aluminium strip with the film outside through 180 °C over a rod or mandrel of specified diameter. The film at the bend is examined for signs of cracks and loss of adhesion. The results will depend on the thickness and surface preparation of the strip, diameter of the mandrel, film thickness and rate of bending.

(b) *Adhesion.* Adhesion is the degree of attachment between a paint or varnish film and the underlying material with which it is in contact. The latter may be another film of paint (adhesion between one coat and another) or any other material such as wood, metal, plaster, etc. (adhesion between a coat of paint and its substrate) (BS 2015).

There is no simple method which satisfactorily measures the adhesion of films to substrates. The *Mandrel test* can be used as a qualitative or comparative method. The *Hounsfield Tensometer* can also be used to measure the force required to pull apart two cylinders glued together by paint. The force is a measure of adhesion.

(c) *Hardness.* This is the ability of a paint film, as distinct from its substrate, to resist indentation or penetration by a hard object (BS 2015).

Hardness of a paint film is developed as a result of solvent evaporation and/or polymerisation process. An absolute method of measurement is not possible. A very simple qualitative assessment of hardness of film is by the thumbnail test. A more accurate method is the *scratch test.* This involves lowering a weighted needle gently on to the painted surface of the metal panel. By smartly pulling the panel, the weighted needle is made to move along the paint film. The weight on the needle is adjusted until the first visible scratch showing the bare metal is obtained. The required weight is a measure of film hardness.

(d) *Opacity* (*or hiding power*). This is quantitatively defined as the extent to which a paint obliterates the colour of an underlying surface of a different colour when a film of it is applied by some standard method (BS 2015).

Assessment of opacity can be obtained by use of *Morest Hiding Power Charts* which have contrasting patterns of black and white. The amount of paint by brush application required to obliterate the contrast pattern is the measure of opacity.

(e) *Film thickness.*

(i) *Wet films.* Film thickness can be determined by placing a plano-convex lens on the film of paint and measuring the diameter of the paint patch left on the underside of the lens. The diameter of the patch is proportional to the thickness of film. A more accurate method makes use of the *wet film thickness gauge* which is based on the same principle.

(*ii*) *Dried films.* Two methods can be distinguished:

I. *Non-destructive method* (*Elcometer*). The principle involves the measurement of magnetic flux and the influence of an applied paint film on the metal. The influence on magnetic flux is proportional to the thickness of the film which acts as an air gap. The Elcometer is the instrument which makes use of this principle and the scale is calibrated to give a direct reading of the film thickness.

It is suitable for non-magnetic paints on ferrous substrates.

II. *Destructive method* (*Rossman Dial Thickness Gauge*). A small patch of paint is removed from the substrate. By resting the two legs of the gauge on the painted areas across the bare patch, the spring-loaded probe is made to rest on the exposed patch and the dial reads directly the film thickness (analogous in principle to a spherometer).

(*f*) *Water resistance.* Several tests are available.

(*i*) *Humidity tests—resistance to blistering.* Test panels are stored in a humidity cabinet under controlled conditions of temperature and humidity. The time taken for blisters to form and the number of blisters developed are recorded and taken as a measure of resistance.

(*ii*) *Water immersion tests.* Test panels are immersed in water at 10–20 °C. Periodically, the panels are examined for the presence and number of blisters, loss of adhesion, loss of gloss or colour and appearance of corrosion.

(*g*) *Corrosion tests* (*resistance to continuous salt spray*). Mild steel panels are coated with primers, allowed to dry for seven days and placed in the salt spray cabinet. The test panels are examined periodically for signs of corrosion.

(*h*) *Durability tests.*

(*i*) *Natural weathering tests.* Test panels are exposed to the weather (sunlight and moisture) and left for a long period of time.

Observations are recorded periodically for any changes such as chalking, colour changes, loss of adhesion, presence of corrosion etc.

This is a very slow process and requires at least three years before any satisfactory assessment can be made.

(*ii*) *Accelerated weathering tests.* Test panels are placed in a weatherometer and subjected to continuous UV radiation from a 900-watt carbon-arc lamp and periodic spraying with distilled water. The panels are observed at regular intervals and any changes taking place are noted.

It is not easy, however, to correlate the results from natural and from accelerated weathering.

Methods of test for paints are dealt with in BS 3900.

PROGRESS TEST 10

1. Explain the difference between the terms paint, varnish, lacquer. (p. 200)

2. By means of a schematic diagram, show the composition of a typical paint. (Fig. 37)

3. What is the purpose of a primer in a paint coat system? Give a typical formulation of a primer for (a) wood, (b) metal. (3, Table LVIII)

4. What is the purpose of an undercoat in a paint coat system? Give a typical formulation of an undercoat. (4, Table LIX)

5. What is the purpose of a finishing coat in a paint coat system? Give a typical formulation of a finishing coat. (5, Table LX)

6. What are the main ingredients used in some special paints in order to suppress fungal and marine growths? (6)

7. Why is it necessary to prepare or pretreat the surface of a substrate prior to painting?
Outline the procedures involved in the surface preparation of metals, timber products and brickwork. (8)

8. Give an outline account of the various methods of paint application. (9)

9. What are the factors which may cause paint defects or failures? (p. 215)

10. What are the paint defects or failures commonly encountered? State the cause of such failures and suggest how they can be prevented or remedied. (10–17)

11. Describe the main tests commonly carried out on (a) liquid paint, (b) dried film. (18–19)

EXAMINATION QUESTIONS

1. (a) An alkyd gloss finishing paint has the following composition:

Pigment:	Titanium dioxide	22%
Extender:	Barium sulphate	5%
Resin:	65% *oil-length soya bean/pentaerythritol alkyd* at 75% solids in white spirit	60%
Driers:	Lead naphthenate	2%
	Cobalt naphthenate	2%
Solvent:	White spirit	9%

(*b*) Explain the functions of the terms in italics. Write an account of the defects that may arise after the application of a paint film due to preparation, selection of materials and workmanship.

(P.S.B. H.N.D. Bldg.)

2. What are the main functions of the following component parts of a paint system:

(*i*) priming coat,

(*ii*) undercoat, and

(*iii*) finishing coat.

Name *one* example of each of the above parts of a paint system for use on

(*i*) wood surfaces,

(*ii*) metal surfaces, and

(*iii*) concrete surfaces, respectively.

Give reasons for your choice.

(P.S.B. H.N.D. S.E.)

3. (*a*) With reference to paint technology, explain the following terms:

(*i*) Newtonian and non-Newtonian fluids,

(*ii*) thixotropy and viscosity.

(*b*) What are the functions of a primary coat of paint? Discuss the preparation and treatment of the following surfaces prior to priming:

(*i*) wood,

(*ii*) concrete,

(*iii*) aluminium.

Suggest one appropriate primer for each type of surface.

(S.O.E. Grad. Exam. C.E.)

4. (*a*) *Brushing*, *spraying* and *dipping* are the three methods of paint application which are widely used. Make a comparative assessment of these methods, in terms of techniques, suitability of objects to be painted and the defects, if any.

(*b*) Discuss the health hazards involved in using the paint spray and the necessary safety precautions to be observed.

(P.S.B. B.Sc. S.E.)

5. A newly-plastered wall is to be painted. Describe the defects that might arise and indicate their causes. What precautions can be taken to minimise their incidence?

(P.S.B. B.Sc. Bldg.)

6. The protective qualities and finished appearance of painted surfaces are dependent upon the preparation of the surface, the use of the correct paints and the correct build-up of coats. Identify the particular problems and prepare brief specification notes for the application of paint to the following surfaces:

(*a*) mild steel structural frame members,

(*b*) external joinery,

(*c*) recently-completed surfaces of Class B gypsum plaster.

(I.O.B. Ass. Part 1 Specimen)

7. (*a*) Describe and give the causes of the following paint film defects:

    (*i*) blistering,

    (*ii*) chalking,

    (*iii*) cracking,

    (*iv*) flaking.

What precautions should be taken to prevent or lessen such defects?

    (*b*) Outline the principle and method of test for the following:

        (*i*) thickness of film,

        (*ii*) hardness of film.

(P.S.B. B.Sc. S.E.)

FURTHER READING

O.C.C.A., *Paint Technology Manuals* (6 parts), Chapman and Hall, 1965–9.

Tatton, W. H. and Drew, E. W., *Industrial Paint Application*, Newnes-Butterworth 1971.

Hess, M., *Paint Film Defects*, Chapman and Hall, 1965.

Chatfield, H. W., *The Science of Surface Coatings*, Ernest Benn, 1962.

CHAPTER XI

METALLIC CORROSION

THE MEANING OF CORROSION

Corrosion, in general, is the destruction of a material by chemical attack resulting from its environment. For metals, it involves a gradual reversion to the more stable state such as the oxide, sulphide or carbonate.

CORROSION MECHANISMS

The mechanisms of corrosion fall into two main types.

1. Direct chemical combination. Here, metals combine directly with gases such as oxygen, chlorine, sulphur gases, carbon dioxide, etc. to form a surface film. With oxygen, a metallic oxide is formed even at normal temperatures and in the absence of moisture:

$$\underset{\substack{\text{(bivalent)}\\\text{metal}}}{2M} + \underset{\text{oxygen}}{O_2} \longrightarrow \underset{\text{oxide}}{2MO}$$

Metals also combine with other gases to form chlorides, sulphides, carbonates, etc., although in general the conversion can take place only at high temperatures.

This type of corrosion is more serious in highly-polluted atmospheres, although in a few cases the corrosion can be self-suppressed by the formation of a thin protective layer of surface film (*e.g.* in aluminium and chromium).

2. Electrochemical corrosion. This is the corrosion in aqueous environments—the result of electrochemical reactions in which the water acts as a conducting liquid (electrolyte).

In an aqueous solution, a metal ionises thus:

$$\underset{\substack{\text{(bivalent)}\\\text{metal}}}{M} \longrightarrow \underset{\text{ions}}{M^{2+}} + \underset{\substack{\text{electrons (negatively}\\\text{charged)}}}{2e^-}$$

The metal ions go into solution and the electrons are left behind in the anode. If the anode is connected with the cathode by

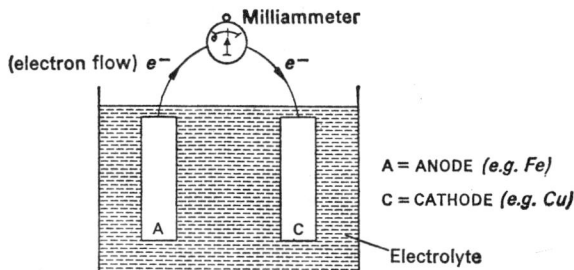

FIG. 38.—*Voltaic or galvanic cell.*

means of a conducting wire, the electrons will flow from the anode along the wire to the cathode (*see* Fig. 38).

(*a*) *Anode reactions* (*oxidation*).

$$Fe \longrightarrow Fe^{2+} + 2e^- \qquad \dots (1)$$

Iron   Ferrous ions

(*b*) *Cathode reactions* (*depending on pH of the medium*).

(*i*) *Acidic medium* ($pH < 7$):

$$2H^+ + 2e^- \longrightarrow H_2 \qquad \dots (2)$$

Hydrogen ions   Hydrogen gas evolved

The hydrogen gas may accumulate on the cathode surface (especially in a neutral electrolyte, *e.g.* sodium chloride (*NaCl*) solution, causing *cathodic polarisation*).

(*ii*) *Alkaline or neutral medium* ($pH \geqslant 7$). Dissolved oxygen in electrolyte may act as a depolarising agent by combining with hydrogen:

$$4H^+ + O_2 + 4e^- \longrightarrow 2H_2O \qquad \dots (3)$$

$$O_2 + 2H_2O + 2e^- \longrightarrow H_2O_2 + 2OH^- \qquad \dots (4)$$

Hydrogen peroxide   Hydroxyl ions

$$O_2 + 2H_2O + 4e^- \longrightarrow 4OH^- \qquad \dots (5)$$

(*c*) *Side reactions* (*in solution*). Finally, the metal ions dissolved from the anode react with the hydroxyl ions, followed by oxidation with the dissolved oxygen.

$$Fe^{2+} + 2OH^- \longrightarrow Fe(OH)_2 \qquad \dots (6)$$

$$2Fe(OH)_2 + \tfrac{1}{2}O_2 + H_2O \longrightarrow 2Fe(OH)_3 \qquad \dots (7)$$

Ferrous hydroxide   Ferric hydroxide (rust)

ELECTROCHEMICAL AND GALVANIC SERIES

The ease with which a metal corrodes depends upon its electronegativity and its standard electrode potential. Electronegativity is the capacity for attracting electrons. The halogens (non-metallic elements) are the most electronegative group of elements, while the alkali metals (Lithium, Sodium, Potassium), are the least electronegative group and also chemically highly reactive. If a metal is immersed in a solution of its ions, a potential difference (electrode potential) is set up between the metal and the solution. This electrode potential depends on the electronegativity of the metal and the ionic concentration of the solution. The metallic electrode potentials are measured with the metal immersed in a normal solution of its ions against the standard Hydrogen electrode acting as reference. By arranging metals in order of their electrode potentials, an electrochemical series is obtained (*see* Table LXVI).

TABLE LXVI: ELECTROCHEMICAL (OR ELECTROMOTIVE FORCE) SERIES

Electrode reaction	*Standard electrode potential at 25 °C (E^0 volt)*	
$K = K^+ + e^-$	$-2\cdot93$	Anodic or more
$Ca = Ca^{2+} + 2e^-$	$-2\cdot87$	basic
$Na = Na^+ + e^-$	$-2\cdot71$	(corroded end)
$Mg = Mg^{2+} + 2e^-$	$-2\cdot37$	
$Al = Al^{3+} + 3e^-$	$-1\cdot66$	
$Zn = Zn^{2+} + 2e^-$	$-0\cdot76$	
$Cr = Cr^{3+} + 3e^-$	$-0\cdot71$	
$Fe = Fe^{2+} + 2e^-$	$-0\cdot44$	
$Cd = Cd^{2+} + 2e^-$	$-0\cdot40$	
$Ni = Ni^{2+} + 2e^-$	$-0\cdot25$	
$Sn = Sn^{2+} + 2e^-$	$-0\cdot14$	
$Pb = Pb^{2+} + 2e^-$	$-0\cdot13$	
$H_2 = 2H^+ + 2e^-$	$0\cdot00$	
$Cu = Cu^{2+} + 2e^-$	$+0\cdot34$	
$2Hg = Hg_2{}^{2+} + 2e^-$	$+0\cdot799$	
$Ag = Ag^+ + e^-$	$+0\cdot80$	
$Hg = Hg^{2+} + 2e^-$	$+0\cdot85$	Cathodic or more
$Au = Au^{3+} + 3e^-$	$+1\cdot50$	noble (protected
$Au = Au^+ + e^-$	$+1\cdot68$	end)

The metals are arranged in decreasing order of negativity and reactivity with potassium at the top and gold at the bottom of the series. By coupling any metal to the one below it in this series, the former is likely to form the anodic (corroded) electrode. For example by coupling iron ($E^0 = -0.44$ volt) to copper ($E^0 = +0.34$ volt) the iron will form the anode and therefore will be corroded, the copper will be the cathode and therefore protected. The electromotive force (*e.m.f.*) set up between them will be $0.34 - (-0.44) = 0.78$ volt.

This electrochemical (or *e.m.f.*) series, although useful to electrochemists, is not practical in metallic corrosion problems where the electrolytic environments differ and where alloys are also considered. A more practical table, known as the

TABLE LXVII: GALVANIC SERIES
IN SEA WATER

Mg	(Anodic)
Mg alloys	
Zn	
Al	
Al alloys	
Cd	
Steel	
Pb–Sn solders	
Sn	
Brasses	
Cu	
Bronzes	
Cu–Ni alloys	
Ti	
Stainless Steel	
Ag	
Au	(Cathodic)

Galvanic series, is based on experimentation with combinations of metals in a variety of environments.

Table LXVII shows the galvanic series of metals and alloys in sea water.

The electrode potentials are dependent partly on the environmental conditions to which the metals are exposed, so in some environments the order of the metals and alloys in the galvanic series may be different from that given above.

FACTORS INFLUENCING CORROSION

Oxygen and moisture (water) are essential for corrosion to occur, but under certain circumstances corrosion can take place in the absence of oxygen (anaerobic corrosion).

Most corrosion mechanisms are electrochemical in nature. Electrochemical corrosion involves the presence of:

(a) Anode (the corroded electrode) ⎫
(b) Cathode (the protected electrode) ⎬ in contact.
(c) Electrolyte (the conducting solution). ⎭

The rate at which the metal anode is corroded depends on:

(i) its position in the electrochemical or galvanic series relative to the cathode.

(ii) the electrical conductivity of the electrolyte.

In addition, the rate of corrosion depends on:

(iii) rate of supply and distribution of oxygen.

(iv) differential ionic concentration of the electrolyte.

(v) structural and chemical uniformity of the metal.

Electrochemical corrosion is possible as a result of galvanic cell formation. Some examples of galvanic cells are summarised below (p. 229).

CORROSION OF FERROUS METALS AND ALLOYS

The corrosion resistance of iron is low compared with other metals. The protective action of oxide (rust) on iron is considerably weaker than the corrosion products of lead, aluminium, tin or zinc. The anodic and cathodic reactions are described on pp. 224–5.

Steels have varying corrosion resistance depending on their composition. Small additions of copper can reduce corrosion under atmospheric conditions, alloying with chromium (about 3%) can reduce the corrosion of steel in sea water. Stainless steels (containing 12–30% Cr and up to 11% Ni) have high resistance to corrosive environments, due to the formation of a very thin protective film of chromium oxide on the metal surface. They are, however, susceptible to inter-granular attack and stress-corrosion cracking.

	Anode *(oxidation)* *Baser phase*	*Cathode* *(reduction)* *Nobler phase*
Dissimilar metals in *contact*		
Zinc (Zn) coupled with iron (Fe)	Zn	Fe
Iron (Fe) coupled with copper (Cu)	Fe	Cu
Microstructure		
Pearlite	α	Carbide
	Lower concen- *tration*	*Higher concen-* *tration*
Differential aeration		
Oxidation	Low oxygen	High oxygen
Dirt or scale	Covered areas	Clean exposed areas
Differential ionic *concentration*		
Electrolyte	Dilute solution	Concentrated solution
	Higher energy	*Lower energy*
Stress cells		
Boundaries	Boundaries	Grain
Grain size	Fine-grain	Coarse-grain
Imperfections	Defect	Perfect
Strains	Cold-worked	Annealed
Stresses	Loaded areas	Non-loaded areas

CORROSION OF NON-FERROUS METALS

3. Aluminium and its alloys. Aluminium, a relatively base metal in the *e.m.f.* series, forms in presence of moist air a superficial adherent film of oxide which protects the metal from further corrosion. This protective oxide film can be removed by chemicals such as strong alkalis, strong acids, inorganic salt solutions, mercury compounds, etc. and by alternate wetting and drying processes. This oxide film is resistant, however, to many organic acids, organic compounds, petroleum products, foodstuffs, alcohols, etc. Corrosion can be reduced by appropriate alloying, by proper design of the compound, by use of a suitable coating or by use of specific inhibitors. Some aluminium alloys are resistant to sea water but high-strength copper-containing alloys must not be used in contact with sea water, or with concrete, mortar, or plaster. Serious corrosion occurs when aluminium alloys are in contact with magnesium oxychloride (used in floor coverings). They should be insulated by a bituminous material.

Corrosion resistance of aluminium alloy can be improved by cladding it with a coating of pure aluminium (Alclad alloy).

4. Copper and its alloys. Copper occurs near the noble end of the *e.m.f.* series and is therefore not a very reactive metal. Its corrosion resistance is expected to be relatively good, but can be greatly improved by alloying.

Copper is attacked by caustic alkalis, ammonia and its compounds, oxidising salts, halogens and sulphides. Some brasses ($Cu–Zn$ alloys) undergo dezincification, whereby zinc is lost causing a general weakening of the metal.

5. Lead and its alloys. Lead, on exposure to polluted environments, forms protective films on the surface of the metal. The corrosion resistance of these films depends on their solubility—for example, the insoluble lead sulphate ($PbSO_4$) can provide good insulation and protection from further corrosive attack, whereas the more soluble lead nitrate ($Pb(NO_3)_2$) is easily attacked by a corrosive environment and corrosion of the metal can then proceed unimpeded. Lead carbonate film ($PbCO_3$) and red lead (Pb_3O_4) can provide some degree of protection, but lead oxide (PbO) is not very protective, especially in soft waters free from dissolved carbon dioxide. Lead dissolves in soft waters and in underground waters containing weak organic acids. Lead is also corroded when in contact with oak or teak and with lime mortars or cement in damp conditions.

Corrosion of lead can be reduced by use of inhibitors (*e.g.* cobalt sulphate in sulphuric acid), by cathodic protection, or by alloying (*e.g.* lead–tin alloy on steel–"Terne-plate").

Anodes of lead and lead alloys are commonly used in cathodic protection systems.

6. Magnesium and its alloys. Magnesium is the lightest of all the constructional metals. Its position near the anodic end of the *e.m.f.* series indicates that it is highly reactive chemically and therefore has low corrosion resistance.

It is attacked by mineral and organic acids, fresh water and salt solutions, but is not attacked by alkalis owing to the formation of a protective film on the surface of the metal.

Magnesium is commonly used as a sacrificial anode in cathodic protection.

Magnesium alloys are susceptible to inter-crystalline corrosion and corrosion-cracking due to presence of internal stress.

To improve corrosion resistance, some form of surface coating is necessary—*e.g.* by oxidation or passivation, by applying coatings of paint, plastics or stoving enamels.

7. Tin and its alloys. Atmospheric corrosion of tin is slow. Protective films are formed when in contact with phosphoric and chromic acids. However, it is attacked by nitric acid and by alkalis. Tin is commonly used as a protective coating on more corrosive metals such as iron, and on copper and brass.

Tin alloys, such as solders, undergo corrosion in some supply waters and the rate of corrosion increases with temperature and any increase in water softness.

8. Zinc and its alloys. Zinc and its alloys have good atmospheric corrosion resistance, due to the formation of protective films of oxides, hydroxides and basic carbonate. Zinc is resistant to most natural waters but is attacked by acids and alkalis. Some passivation can be obtained by immersing zinc in chromate solution, due to the formation of a protective film of zinc chromate.

Zinc is widely used as zinc coating in paints (*e.g.* zinc-rich primers) and as the sacrificial anode in cathodic protection.

PROTECTION OF METALS AGAINST CORROSION

Corrosion processes can be varied and complex. A good understanding of their mechanisms can be useful in reducing and minimising their destructive effect on metals. The following are some of the preventive measures which may be taken.

9. By design. Attention to design is the first step to be taken towards corrosion prevention. Detail must be such as to prevent the formation of a galvanic cell, by avoiding:

(a) bi-metallic contact with corrosive medium—insulate the two dissimilar metals from each other;

(b) formation of crevices and moisture traps, inadequate drainage—improve design features (see Fig. 39(a));

(c) bad ventilation—improve drainage and ventilation;

(d) contact with stray electric currents (see Fig. 39(b));

(e) formation of internal stress—avoid sharp corners and sharp bends (see Fig. 39(c));

(f) reinforced concrete: inadequate cover of steel—ensure that at least the steel reinforcements are embedded in a layer (at least 38 mm thick) of a dense concrete mix.

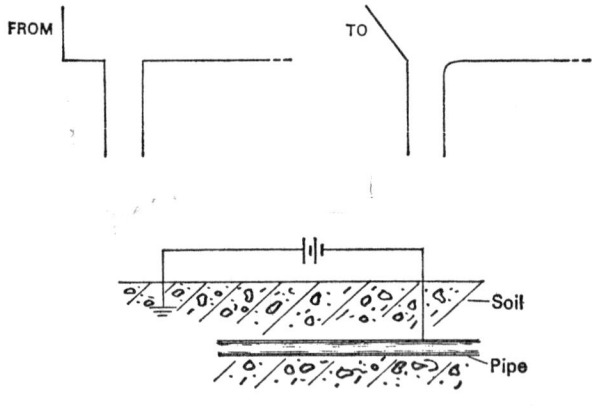

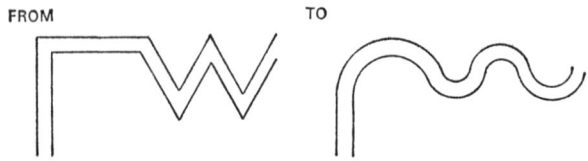

Fig. 39.—*Protection against corrosion: design features.*

(a) (*top*) *Avoidance of crevices and moisture traps.*
(b) (*centre*) *Avoidance of contact with stray electric currents.*
(c) (*bottom*) *Avoidance of internal stress.*

10. By adding suitable alloying elements. In order to increase resistance to corrosive environments, *e.g.* copper in ferritic steel, chromium (12–20 per cent) in stainless steels; copper, phosphorus, silicon in weathering steels; non-ferrous metals and alloys.

11. By treatment of corrosive environment.

(*a*) Removal of polluted atmospheres and reduction of atmospheric humidity, for example: air conditioning in needle factories and dehumidisation in ship holds.

(*b*) Addition of corrosion inhibitors, such as potassium dichromate, sodium nitrate, sodium benzoate.

(*c*) Deaeration of boiler water, for example by addition of sodium sulphite to remove dissolved oxygen.

$$\underset{\substack{\text{sodium}\\\text{sulphite}}}{2Na_2SO_3} + \underset{\substack{\text{dissolved}\\\text{oxygen}}}{O_2} \longrightarrow \underset{\substack{\text{sodium}\\\text{sulphate}}}{2Na_2SO_4}$$

12. Cathodic protection. This involves protecting the metal from corrosion by making it cathodic. This can be achieved by:

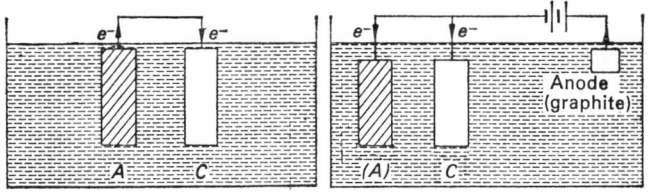

(*a*) *Application of an external D.C. source.*

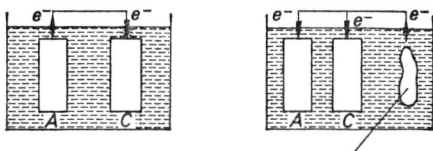

Sacrificial metal

(*b*) *Sacrificial corrosion.*

Fig. 40.—*Cathodic protection against corrosion.*

(*a*) Applying an external D.C. source (*see* Fig. 40(*a*)). An external D.C. source may be applied to graphite or platinum-clad tantalum anodes.

(*b*) Using a baser metal (sacrificial corrosion) (*see* Fig. 40(*b*)).

For cathodic protection of iron, the sacrificial anode can be magnesium, zinc and aluminium, all of which are baser than iron (*see* electrochemical series, p. 226). Owing to their low cost and ease of replacement, magnesium and zinc anodes are often used sacrificially to protect pipes, tanks and ships' hulls. Magnesium alloy containing 6% Al, 3% Zn and 0·2% Mn is very widely used as a sacrificial anode.

Anodic protection of iron is also possible. This involves increasing the potential of the corroding anode, using an external D.C. source so as to give passivity to the corroding metal. The process can be very efficient but requires very careful control, because increased potential can lead to very rapid corrosion under conditions where passivity of the metal does not occur.

13. By use of protective coatings. Corrosion of a metal can be suppressed by covering it with a layer of (*a*) a non-corrodible metal (metallic coating), (*b*) an inorganic coating, or (*c*) an organic coating. The efficiency of such a protective coating depends very largely on the correct pretreatment of the metal surfaces—which involves weathering or mechanical treatment, pickling and degreasing.

(*a*) *Metallic coatings.* Such coatings may be applied by (*i*) dipping, (*ii*) electroplating, (*iii*) spraying, (*iv*) cladding, (*v*) cementation methods.

(*i*) *Dipping.* This is carried out by dipping the metal to be coated in a melt of another metal, usually with low melting point, *e.g.* tin, lead, aluminium or zinc. Some common examples are zinc-coating on iron or steel (galvanised steel), chromium-coating on nickel (chrome-plating), tin on steel, copper and brass (tinning), lead on steel (Terne Plate), aluminium on steel.

(*ii*) *Electroplating.* A thinner coating can be applied to the metal or non-metal by an electrochemical process known as electroplating. This requires a cathode (metal or non-metal to be plated), an anode (the plating metal), an electrolyte and an applied potential from a D.C. source. The mechanism takes place at the cathode as follows:

$$M^{2+} + 2e^- \longrightarrow 2M$$

metal ions → metal (electro-plating at cathode)

The electrolyte is normally an aqueous solution of a salt of the plating metal, often in presence of some addition agents to improve the working of the process.

Copper- or nickel-plating on steel affords good protection against corrosion. Zinc-plating on steel, cadmium-plating, tin-plating, lead-plating and chromium-plating are some examples.

(*iii*) *Spraying.* Molten metal (such as aluminium, zinc, tin, copper, lead, brass or bronze) is sprayed from a special spray gun on to the base metal or non-metal. The coating metal, in the form of wire or powder, is melted electrically or by oxyacetylene flame and the molten metal is dispersed by compressed air.

(*iv*) *Cladding.* The coating metal (*e.g.* copper, nickel, aluminium) is cast around the base metal (*e.g.* steel), followed by heating and rolling to produce a cladding of any desired thickness. Superficial alloying occurs at the surface. This does not occur in the processes described in (*i*)–(*iii*) above.

(*v*) *Cementation methods.* The base metal (*e.g.* mild steel) is coated with a powder of the coating metal (*e.g.* zinc, aluminium) and then strongly heated, when alloying of the two metals occurs. The various processes are *sherardising* (coating with zinc), *calorising* (coating with aluminium) and *chromising* (coating with chromium).

(*b*) *Inorganic coatings.* Some metals, such as aluminium and chromium, do naturally develop a very thin surface protective film of oxide or hydroxide. More resistant and thicker layers of oxide, phosphate and chromate can be formed by chemical treatment.

(*i*) *Oxides.* Oxide films can be formed on aluminium by an electrolytic process known as anodising. This involves the anodic oxidation of aluminium by an applied current in an electrolytic solution of chromic, sulphuric or oxalic acid. The cathode may be stainless steel or lead.

(*ii*) *Chromate.* A thin layer of chromate can be formed on a metal (zinc, aluminium, magnesium, cadmium, copper, bronze) by dipping in a chromic acid or an acid solution of dichromate. Chromate films are generally used as a basis or key for paint and for greater protection it is necessary to follow this treatment by painting with a coat of varnish or paint (such as zinc chromate).

(*iii*) *Phosphate*. The main use of phosphate-coating, like chromate-coating, is to provide a basis or key for paints.

Phosphate films can be formed on metals (such as iron, zinc, cadmium, aluminium) by immersion in a phosphating bath generally containing phosphoric acid and the primary phosphates of magnanese, zinc or iron together with an oxidant. Phosphating treatments are commercially known as "Bonderising," "Parkerising," "Granodising,' "Walterising", etc.

(*c*) *Organic coatings*. This involves the application of paints (including bituminous paints), varnishes, lacquers, plastics and rubbers in order to isolate the metal from the corrosive environment. The efficiency of this method of control depends on adequate surface preparation of the metal, appropriate method of application (*e.g.* brushing, dipping, spraying, etc.), correct choice of paints and compatibility of paint systems (*e.g.* primer, undercoat, finishing coat).

14. By use of inhibitors. An inhibitor is a chemical substance which when added in small quantities to a corrosive environment effectively decreases the corrosion rate at any metal surface with which it is in contact.

Anodic inhibitors. These affect particularly the anodic processes, by forming insoluble compounds (oxides, hydroxides, silicates, borates, phosphates, carbonates, benzoates, chromates and nitrites) at the anode. If sufficient amounts of such inhibitors (except benzoates) are used, severe local attack may occur.

Cathodic inhibitors. These affect the cathodic processes by forming a thin film on the cathode. They are not generally as efficient as anodic inhibitors but they are safer in that local attack is not intensified. They include salts of magnesium, nickel and zinc which form insoluble hydroxides on the cathode. The thin film protects the cathode from oxygen which would otherwise depolarise the cathode and so increase the rate of corrosion.

PROGRESS TEST 11

1. What is meant by "metallic corrosion"? (p. 224).
2. Describe the mechanism of electrochemical corrosion (2).
3. The following pairs of metals are in contact in an electrolyte:

(a) iron and copper, (b) iron and zinc, (c) iron and magnesium, (d) lead and copper.

Using the table of the electrochemical series, state which metal will be corroded and in each case calculate the *e.m.f.* set up (Table LXVI).

4. Why is a galvanic series more practical than an electrochemical series in the study of corrosion problems? (p. 227).

5. What are the factors which may cause corrosion? Give some examples of galvanic cells (p. 228).

6. Why is iron more readily corroded than non-ferrous metals such as aluminium or copper? (3, 4).

7. Describe the various methods that can be used in order to protect metals from corrosion (9–14).

EXAMINATION QUESTIONS

1. (a) State the conditions required for the corrosion of ferrous metals.

(b) Describe, with the aid of a diagram, the mechanism of electrochemical corrosion of ferrous metals.

(c) Discuss briefly how a knowledge of the above mechanism can be used to minimise the corrosion of ferrous structures.

(P.S.B. "A" Level Qualifying Course)

2. Comment on the following statements:

(a) Magnesium bar is often used in water storage tanks.

(b) When two dissimilar metals are coupled together it is necessary to insulate them by plastic insulator.

(c) Rusting of an iron nail takes place quicker in salt water than in freshly-boiled distilled water.

(d) 18:8 stainless steel has a high resistance to corrosion.

(e) Steel sections embedded in concrete are protected against corrosion.

(P.S.B. B.A. Arch.)

3. What are the causes of corrosion commonly encountered in building? Describe briefly, giving one practical example in each case:

(a) Stress corrosion,

(b) Sacrificial corrosion,

(c) Concentration cell.

Suggest in each case one different method of reducing the risk of corrosion.

(P.S.B. Grad. Dip. Arch.)

4. (a) Describe the mechanism of rusting of steel exposed to a neutral atmospheric environment.

(b) Discuss the factors which must be considered before the selection of a corrosion prevention scheme for a steel structure.

(c) Outline a protective scheme, giving reasons for your choice, for a mild steel structure exposed to a severe environment.

(P.S.B. B.Sc. Bldg.)

5. Give the meaning of the terms: "wet" corrosion, "dry" corrosion. What happens when:

(a) steel rivets are coupled with copper plates,

(b) aluminium brackets are in contact with an external concrete wall,

(c) steel structures are exposed to a highly polluted atmosphere?

What precautions, if any, should be taken in order to protect from corrosion?

(S.O.E. Grad. Exam. C.E. Specimen)

6. (a) Describe the concentration cell mechanism of corrosion of steel exposed to an atmospheric environment.

(b) (i) Explain why reinforcing steel is protected against corrosion by the concrete cover.

(ii) Describe the effects of corrosion of reinforcement in concrete.

(iii) Outline the methods available for preventing the corrosion of reinforcement in concrete.

(P.S.B. B.Sc. S.E.)

FURTHER READING

Evans, U. R., *An Introduction to Metallic Corrosion*, Edward Arnold, 1960.

Scully, J. C., *Fundamentals of Corrosion*, Pergamon, 1966.

Diamant, R. M. E., *The Prevention of Corrosion*, Business Books Ltd., 1971.

CHAPTER XII

METALS AND ALLOYS

INTRODUCTION

There are over 100 elements known at present, some eighty of them are metals but only nine of these (*Fe, Cu, Al, Pb, Sn, Mg, Ni, Ti, Zn*) are important engineering materials (*see* Periodic Table, Table LXVIII). Those materials which are based on iron are termed "Ferrous" while those which are based on other metals are called "Non-ferrous." Light metals include aluminium and magnesium, while titanium is one of the newer metals.

EXTRACTION OF METALS

Metals seldom occur in nature in a virgin state but in a chemically combined state (as oxides, sulphides, carbonates, chlorides, etc.) and as ores. Generally, the methods of extracting metals from ores are of two types:

(*a*) *Thermal reduction.*

$$\text{Metal ore} + \underset{(e.g.\ \text{carbon})}{\text{reducing agent}} \xrightarrow{\text{heat}} \text{Metal} + \text{gas}$$

(*b*) *Electrolytic reduction.* Electrolysis of an aqueous solution or a melt of the salt of metal results in a purer form of the metal being obtained at the cathode.

In order to obtain metals of high purity, electrolytic refining is often necessary.

Table LXIX summarises the extraction of the nine common engineering metals, together with their important alloys.

GENERAL PROPERTIES OF METALS

Metals (*see* Table LXX) are crystalline solids of high density and high melting point (except mercury, which is liquid at room temperature). They are good conductors of heat and

TABLE LXVIII: PERIODIC TABLE OF THE ELEMENTS

Group	IA	IIA	IIIA	IVA	VA	VIA	VIIA	VIII			IB	IIB	IIIB	IVB	VB	VIB	VIIB	O
Outer electron configuration (n = period number)	ns^1	ns^2						$(n-1)d^1, ns^2 \longrightarrow (n-1)d^{10}, ns^2$					$ns^2 np^1$	$ns^2 np^2$	$ns^2 np^3$	$ns^2 np^4$	$ns^2 np^5$	$ns^2 np^6$
Period 1	^1H																	^2He
Period 2	^3Li	^4Be											^5B	^6C	^7N	^8O	^9F	^{10}Ne
Period 3	^{11}Na	^{12}Mg											^{13}Al	^{14}Si	^{15}P	^{16}S	^{17}Cl	^{18}Ar
Period 4	^{19}K	^{20}Ca	^{21}Sc	^{22}Ti	^{23}V	^{24}Cr	^{25}Mn	^{26}Fe	^{27}Co	^{28}Ni	^{29}Cu	^{30}Zn	^{31}Ga	^{32}Ge	^{33}As	^{34}Se	^{35}Br	^{36}Kr
Period 5	^{37}Rb	^{38}Sr	^{39}Y	^{40}Zr	^{41}Nb	^{42}Mo	^{43}Tc	^{44}Ru	^{45}Rh	^{46}Pd	^{47}Ag	^{48}Cd	^{49}In	^{50}Sn	^{51}Sb	^{52}Te	^{53}I	^{54}Xe
Period 6	^{55}Cs	^{56}Ba	$^{57-71}*$	^{72}Hf	^{73}Ta	^{74}W	^{75}Re	^{76}Os	^{77}Ir	^{78}Pt	^{79}Au	^{80}Hg	^{81}Tl	^{82}Pb	^{83}Bi	^{84}Po	^{85}At	^{86}Rn
Period 7	^{87}Fr	^{88}Ra	$^{89-103}$†															

* Lanthanides (filling 4f)	^{57}La	^{58}Ce	^{59}Pr	^{60}Nd	^{61}Pm	^{62}Sm	^{63}Eu	^{64}Gd	^{65}Tb	^{66}Dy	^{67}Ho	^{68}Er	^{69}Tm	^{70}Yb	^{71}Lu
† Actinides (filling 5f)	^{89}Ac	^{90}Th	^{91}Pa	^{92}U	^{93}Np	^{94}Pu	^{95}Am	^{96}Cm	^{97}Bk	^{98}Cf	^{99}Es	^{100}Fm	^{101}Md	^{102}No	^{103}Lw

TABLE LXIX: EXTRACTION OF COMMON ENGINEERING METALS

Metal	Ore	Extraction	Alloys
Iron (Fe)	Haematite (Fe_2O_3) Magnetite (Fe_3O_4)	*Thermal reduction* with coke (carbon): (i) $2C + O_2 \rightarrow 2CO$ (reducing agent) (ii) $Fe_3O_4 + 4CO$ $\rightarrow 3Fe + 4CO_2$ followed by purification of crude iron (pig iron)	Steels
Copper (Cu)	Copper pyrites $(CuFeS_2)$ copper glance malachite azurite	(i) *Copper pyrites* $\rightarrow$ "copper concentrate" by flotation process (ii) *Smelting:* $2CuFeS_2$ $+ 2SiO_2 + 4O_2$ $\rightarrow Cu_2S + $ slag $Cu_2S + 2O_2$ $\rightarrow 2CuO + SO_2$ $Cu_2S + 2CuO$ $\rightarrow 4Cu + SO_2$ (iii) *Refining:* fire or electrolytic refining	Brasses $(Cu-Zn)$ Bronzes $(Cu-Sn)$ Al bronzes $(Cu-Al)$ Cupronickels $(Cu-Ni)$
Aluminium (Al)	Bauxite $(Al_2O_3.nH_2O)$	(i) *Reduction:* Bauxite $\rightarrow Al_2O_3$ (ii) *Electrolysis* of a molten solution of alumina in cryolite (Na_3AlF_6) at about 950 °C (iii) *Refining:* Electrolytic refining	Heat-treatable: $Al-Cu$ $Al-Cu-Ni$ $Al-Cu-Zn$ $Al-Mg-Si$ Non-heat-treatable: $Al-Mn$ $Al-Mg$ $Al-Si$
Magnesium (Mg)	Magnesite $(MgCO_3)$ Brucite $(Mg(OH)_2)$ Dolomite $(CaCO_3.MgCO_3)$	*Thermal process:* (i) Calcination $\rightarrow MgO$ (ii) Reduction $\rightarrow Mg$ (iii) Refining by distillation or electrolytic process: converting sea water to $Mg(OH)_2$, then to $MgCl_2$, which is reduced electrolytically to Mg	$Mg-Al-Zn$ $Mg-Zn-Zr$ $Mg-Mn$
Nickel (Ni)	Pentlandite $(NiFe)_9S_8$ Millerite (NiS) Polydymite (N_3S_4)	(i) *Ore concentration:* flotation or magnetic process (ii) *Roasting and smelting* $\rightarrow Ni$ sulphide (iii) *Refining:* electrolytic refining	Monel: $Ni-Cu-Fe$ Inconel: $Ni-Cr-Fe$ Nimonic: $Ni-Cr$ (also $Al-Cu-Ni$, $Cu-Ni$)
Tin (Sn)	Cassiterite (SnO_2)	(i) *Ore concentration:* flotation process (ii) *Thermal reduction* $\rightarrow$ crude tin (iii) *Refining:* fire or electrolytic refining	Solders Bearing alloys Bronze
Lead (Pb)	Galena (PbS) Cerussite $(PbCO_3)$ Anglesite $(PbSO_4)$	(i) *Ore concentration:* by oil-flotation (ii) *Oxidation of galena* $\rightarrow PbO$ (iii) *Thermal reduction:* $PbO + C \rightarrow Pb + CO$ (iv) *Refining:* electrolytic refining	Solders: $Pb-Sn$ $Pb-Sn-Ag$ Bearing alloys: $Pb-Sn-Sb$ Fusible alloys

TABLE LXIX continued

Metal	Ore	Extraction	Alloys
Zinc (*Zn*)	Zinc blende (*ZnS*) usually associated with galena (*PbS*)	(*i*) *Ore concentration:* by oil-flotation (*ii*) *Oxidation: ZnS → ZnO* (*iii*) *Reduction: ZnO + C → Zn* (*iv*) *Refining:* distillation or electrolytic process	*Bearing alloy: Zn–Ag–Cu Brasses: Cu–Zn (also Al–Cu–Zn, Mg–Al–Zn)*
Titanium (*Ti*)	Rutile (*TiO₂*) Ilmenite (*TiFeO₃*)	(*i*) *Ore concentration:* by flotation (*ii*) *Chlorination of rutile concentrate →* crude *TiCl₄* (*iii*) *Reduction to TiCl₄ and finally Ti:* Thermal or electrolytic process	*Ti–Al–(Mn–V)* main uses in aircraft construction

TABLE LXX: PROPERTIES OF

Metal	Melting point (°C)	Boiling point (°C)	Crystal structure[1]	Density (kg/m³)	Young's modulus (kN/mm²)	Electrical resistivity ($\times 10^{-8}$ Ωm)
Iron (*Fe*)	1535	3235	bcc (α) fcc (γ) bcc (δ)	7860	193	10
Copper (*Cu*)	1083 (99·9% purity)	2575	fcc	8890	122–132	1·724
Aluminium (*Al*)	659	2441	fcc	2700	68·3–72·3	2·845
Magnesium (*Mg*)	649	1090	cph	1740	44	3·9
Nickel (*Ni*)	1453	2800	fcc	8910	210	6·844
Tin (*Sn*)	232 (β)	2620 (β)	bct (white or β) bcc (grey or α)	7290 (β) 5770 (α)	40·8	11·5–12·8
Lead (*Pb*)	327·4	1750	fcc	11344	13·8–16·5	20·63
Zinc (*Zn*)	419·5	905·7	cph	7130	90	5·9
Titanium (*Ti*)	1668	3260 3535	cph (α) bcc (β)	4510	106	47·8

[1] Crystal structure: bcc = body-centred cubic, fcc = face-centred cubic, cph = close-packed hexagonal, bct = body-centred tetragonal.

electricity through the nature of their inter-atomic bonding—
see the nature and properties of metallic bonding in Table
LXXI.

IRON AND STEEL

The common iron ore is either magnetite (Fe_3O_4) or haematite (Fe_2O_3) which contains appreciable amounts of impurities
(as SiO_2, CaO, Al_2O_3) together with traces of sulphur and
phosphorus. Thermal reduction is carried out in a blastfurnace
using coke as the reducing agent and limestone ($CaCO_3$) as a
fluxing agent to remove silica (SiO_2). The chemical process

COMMON ENGINEERING METALS

Thermal conductivity (W/mK)	Coefficient of linear expansion ($\times 10^{-6} K^{-1}$)	Specific heat capacity (J/gK)	Tensile strength (N/mm²)	Elongation on 50 mm length (%)	Hardness HB
71	11·9	0·46	423–510 (mild steel)	22 (mild steel)	130 (mild steel)
393	17	0·385	155–165 310–385 216–247	25–30 5–20 50–60	40–45 as cast 80–115 cold-worked 45–55 annealed after cold working
218·5	24·0	0·925	62 (annealed) 102 (hard)	45 (annealed) 6 (hard)	15 (extrusions) 15–30 (sheet)
146	29·9	1·04	170 (UTS) (annealed)	5 (annealed)	40 (annealed)
83	13·3	0·512	370–400 (annealed bar)	40–50	80–110 (annealed bar)
64	20	0·225	15	75	4
35	29·5	0·130	11·1 (UTS)	68·6	4
113	33–39·5	0·393	200–500 (annealed)	27 (annealed)	45–50
17	8·5	0·58	300–450	30 (annealed)	

TABLE LXXI: NATURE AND PROPERTIES OF INTER-ATOMIC BONDING

Type of bonding	Nature	General properties of compounds
ELECTRO-VALENCY	Exists only in compounds Involves electron transfer $Na \quad + \quad \cdot \ddot{\underset{\cdot\cdot}{Cl}} : \quad \longrightarrow \quad Na^+ \quad + \quad \left[\ddot{\underset{\cdot\cdot}{O\,Cl}} : \right]^-$ $1s^2 2s^2 2p^6 3s^1 \quad 1s^2 2s^2 2p^6 3s^2 3p^5 \qquad 1s^2 2s^2 2p^6 \qquad 1s^2 2s^2 2p^6 3s^2 3p^6$	Strong and hard crystals High melting point and high boiling point Insulators; conduction in the melt by ion-transport Non-directional Readily soluble in water; only sparingly soluble in organic solvents
COVALENCY	Exists in elements (non-metallic) and in compounds Involves electron sharing $H \cdot + \cdot H \longrightarrow H : H \ (\text{or } H — H)$ $1s^1 \quad 1s^1$	Strong and hard crystals Lower melting point and boiling point than electrovalent compounds Insulators in the solid and in the molten state Strongly directional More readily soluble in organic solvents

METALLIC	Exists in metallic elements and compounds Attraction between metal ions and the electron cloud; electrons are usually non-localised — = Electron cloud	Variable strength and hardness Variable melting point Conduction of electric current (by electron transport) Non-directional
VAN DER WAALS'	Arising from the interactions of permanent and of induced dipoles $\delta+$ $\delta-$ $\delta+$ $\delta-$ $H----Br--------H--------Br$	Weak and soft crystals Low melting point Insulators Non-directional
HYDROGEN BONDING	A dipole–dipole attraction between a hydrogen atom attached to a strongly electronegative atom and a second strongly electronegative atom with a lone pair of electrons	Rather similar to Van der Waals'

taking place can be represented by the following chemical reactions:

(i) $2C + O_2 \longrightarrow 2CO$
coke air   carbon monoxide
                (reducing agent)

(ii) $Fe_3O_4 + 4CO \longrightarrow 3Fe + 4CO_2$
magnetite   pig   carbon
                             iron   dioxide

or

$Fe_2O_3 + 3CO \longrightarrow 2Fe + 3CO_2$
haematite

(iii) $CO_2 + C \longrightarrow 2CO$

(iv) $CaCO_3 + SiO_2 \longrightarrow CaSiO_3 + CO_2$
limestone   silica   calcium silicate
            (impurities)   (slag)

The resulting metal (pig iron) is rather impure, containing about 3·5 per cent carbon together with sulphur, silicon, phosphorus and manganese. These impurities can be removed *either* by passing air or pure oxygen, when the following reactions take place:

$2C + O_2 \longrightarrow 2CO$
$C + O_2 \longrightarrow CO_2$
$S + O_2 \longrightarrow SO_2$
$Si + O_2 \longrightarrow SiO_2$

or by adding iron oxide to melt:

$5FeO + 2P \longrightarrow P_2O_5 + 5Fe$
                phosphorus pentoxide

$FeO + C \longrightarrow CO + Fe$

The purified pig iron now becomes *wrought iron* (with a carbon content of about 0·02 per cent). Controlled amounts of carbon together with manganese, chromium, nickel, titanium, molybdenum, vanadium and tungsten can then be added to produce steel which can be considered as an iron–carbon alloy. The various processes used for the manufacture of steel include Bessemer, open-hearth, electric arc, LD (Linz-Donawitz), Kaldo, etc.

1. Fe–C alloys. The main ones include wrought iron (0·02 per cent C + slag), steel (0·1–1·7 per cent C), and cast iron (2·4–4 per cent C).

(a) *Wrought iron* is characterised by low carbon content (0·02 per cent C), high ductility (25–40 per cent elongation at failure), moderately high strength (ultimate tensile strength 340–360 N/mm^2), good hardness (Brinell hardness number 75–100) and relatively good corrosion resistance (often better corrosion resistance than mild steel). It is used for gates, fencing, crane hooks, chains, anchors, railway couplings, etc. The chemical composition and properties of wrought iron vary according to types—*see* typical composition and properties in Table LXXII.

TABLE LXXII: TYPICAL COMPOSITION AND PROPERTIES
OF WROUGHT IRON AND PIG IRON

	COMPOSITION (%)					PROPERTIES		
	C	Si	S	P	Mn	Yield point (N/mm^2)	Tensile strength (N/mm^2)	Elonga- tion (%)
Wrought iron (Staffordshire)	0·02	0·12	0·018	0·228	0·01	230	355	25
Pig iron	3·5	1·9	0·06	1·0	0·7	NIL	355	NIL

[*Rollason, E. C., Metallurgy for Engineers, Edward Arnold, 1973.*]

(b) *Cast iron* is obtained by remelting selected grades of pig iron, adjusting its composition and then casting in sand moulds. It contains 2–4 per cent carbon which exists either as an unstable iron carbide (Fe_3C), known as cementite, or as free carbon (graphite).

Various types of cast iron are available, which vary in composition and properties:

(*i*) *White cast iron.* Most of the carbon is present as cementite, which makes it very hard and brittle and hence non-machinable. Its use is therefore limited, mainly as a raw material in the manufacture of malleable cast irons.

(*ii*) *Grey cast iron.* This is readily machinable, owing to the precipitation of carbon in the form of graphite from cementite. Free graphite exists either as flakes or in nodular form as spheroidal–graphite iron (SG iron).

Grey cast iron castings are specified according to BS 1452: 1961 (amended 1969) which gives seven grades (*see* Table LXXIII).

Typical applications include:

Automobile engineering:	brake drums and discs, crankshafts, cylinder liners and pistons, etc.
Heavy engineering:	marine propellers, gearwheels, turbine vanes, etc.
General engineering:	machinery parts, components for all types of mechanical handling equipment.
Mining and metal production:	blastfurnace parts, dies, moulds, etc.
Building and domestic industries:	baths, cisterns, rainwater pipes and gutters, radiators, boilers, etc.

TABLE LXXIII: PROPERTIES OF GREY CAST IRON

BS 1452	Minimum tensile strength (N/mm²)	Modulus of elasticity (kN/mm²)	Minimum elongation (%)	Hardness (HB)	Specific gravity	Coefficient of thermal expansion (× 10⁻⁶ K⁻¹)
Grade 10	153	76–103	0·5–0·75	160–180	7·0	11–14
Grade 12	185	83–110	0·5–0·75	160–180	7·1	11–14
Grade 14	216	97–117	0·5–0·75	180–240	7·2	11–14
Grade 17	262	110–131	0·5–0·75	190–250	7·3	11–14
Grade 20	309	124–145	0·5–0·75	220–280	7·3	11–14
Grade 23	355	124–145	0·5–0·75	220–300	7·4	11–14
Grade 24	402	124–145	0·5–0·75	240–300	7·5	11–14

(iii) Spheroidal-graphite iron (SG iron). The strength of the cast iron can be improved by altering the nature of the carbon from free flake graphite into the nodular or spheroidal form by inoculation of the melt with small amounts of magnesium or cerium prior to casting. Addition of nickel further improves the properties of SG iron.

BS 2789 : 1973 specifies six grades of SG iron with normal matrix structures ranging from ferritic to pearlitic. The ferritic structure is characterised by easy machinability and high ductility, whereas the pearlitic structure gives rise to higher strength (*see* Table LXXIV).

Typical applications include:

Automobile engineering:	brake and clutch pedals, crankshafts, front axle supports, etc.
Heavy engineering:	casings and cylinders, farm machinery, pressure pipes and fittings, etc.

TABLE LXXIV: TYPICAL PROPERTIES OF SG IRON

BS 2789	Normal heat treatment	Normal matrix structure	Ultimate tensile strength (Minimum) (N/mm²)	Modulus of elasticity (kN/mm²)	Minimum elongation (%)	Hardness (HB)	Specific gravity	Coefficient of thermal expansion (×10⁻⁶K⁻¹)
SNG 24/17	2-4 hr. at 900 °C 6-12 hr. at 700 °C	Fully ferritic	371	172	17	140-170	7·1	11·5
SNG 27/12		Mainly ferritic	417	172	12	140-180	7·1	11·5
SNG 32/7	2-4 hr. at 900 °C, controlled furnace-cool	Ferritic-pearlitic	494	172	7	180-220	7·2	11·5
SNG 37/2	As cast condition	Mainly pearlitic	571	172	2	210-310	7·2	11·5
SNG 42/2	2-4 hr. at 800–900 °C, air-cool	Pearlitic	649	172	2	240-280	7·2	11·5
SNG 47/2	2-4 hr. at 850–900 °C, air-cool or quench and temper	Pearlitic or tempered martensite	726	172	2	250-450	7·2	11·5

Mining and metal production:	blastfurnace parts, coke oven doors and frames, dies, etc.

(*iv*) *Malleable iron.* This is obtained by heat treatment (such as the annealing process) of white cast iron. Three main types are available:

Whiteheart (BS 309 : 1972)
Blackheart (BS 310 : 1972)
Pearlitic (BS 3333 : 1972)

Typical properties of malleable cast iron are summarised in Table LXXV.

Malleable cast irons are used in situations where ductility, machinability and high resistance to atmospheric resistance are called for, *e.g.* in the automobile industry.

2. Steels. Steel is basically an alloy of iron and carbon, together with smaller amounts of other alloying elements such as silicon, manganese, phosphorus, sulphur, nickel, chromium, molybdenum, etc.

Steel can be conveniently classified into two main groups: the plain carbon steel and the alloy steel.

(*a*) *Plain carbon steel.* This can be regarded as steel containing up to 1·5 per cent carbon, not more than 0·5 per cent silicon and 1·5 per cent manganese and only traces of other alloying elements.

It may be subdivided according to the carbon content into four types, as shown in Table LXXVI.

Properties of plain carbon steel depend not only on composition (*i.e.* carbon content) but also on microstructure (*i.e.* size and nature of grains). The finer-grained is normally stronger than the coarser-grained structure. The microstructure is governed by the introduction of alloying elements and/or by heat treatment.

Figure 41 shows graphically the mechanical properties and the microstructure of plain carbon steel.

(*b*) *Alloy steels.* An alloy steel may be regarded as one containing more than 0·5 per cent silicon and 1·5 per cent manganese. Many other alloying elements (*see* Table LXXVII—effect of alloying elements) may be introduced in

TABLE LXXV: TYPICAL PROPERTIES OF MALLEABLE CAST IRON

Types	Grade	Minimum tensile strength (N/mm²)	Minimum 0·5% proof-stress (N/mm²)	Minimum elonga-tion (%)	Modulus of elas-ticity (kN/mm²)	Coefficient of thermal expansion (×10⁻⁶K⁻¹)	Hard-ness (HB)
Whiteheart (BS 309)	W 340/3	270–340	170–220 (0·2% PS)	3–7			120–229
	W 410/4	350–410	190–250	4–10	175·8	10–12·5	
Blackheart (BS 310)	B 290/6	290	170	6			110–149
	B 310/10	310	190	10	168·9	10–12·5	
	B 340/12	340	200	12			
Pearlitic (BS 3333)	P 440/7	440	270	7			149–197
	P 510/4	510	310	4			170–229
	P 540/5	540	340	5	172·4	10–12·5	179–229
	P 570/3	570	420	3			197–241
	P 690/2	690	540	2			241–285

varying amounts and/or in combination with one another in order to improve desirable properties and sometimes long-term economy.

Alloy steels cover such a wide range of alloys that simple classification is not easy—the classification and codings in the British Standards Specifications are too detailed.

TABLE LXXVI: TYPES OF PLAIN CARBON STEEL

Type	Carbon content %	Characteristics	Typical uses
Low carbon steel	up to 0·15	Soft, tough and ductile (since mainly ferrites) Low tensile strength	Wire, rod, tubing, sheet and strip, concrete reinforcing bars
Mild steel	0·15–0·25	Hardness increases with C content Tensile strength increases up to the eutectoid 0·8%C and then decreases only slightly Ductility decreases up to the eutectoid 0·8%C and then decreases only slightly Easily worked and welded	Steel plate and sections (structural steels)
Medium carbon steel	0·20–0·50	Strength and hardness improved at the expense of ductility, can be forged, rolled and machined	Forgings, bright drawn bar, shafts and high tensile tubes, shafts, gears and tyres
High carbon steel	0·50–1·50	Hardest of plain carbon steels High tensile strength, resistance to wear Decrease in ductility and toughness	Forging dies, drills, woodworking and metal-cutting tools

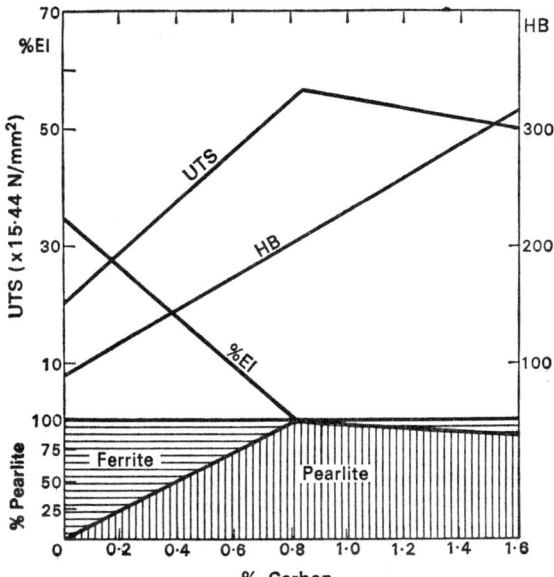

FIG. 41.—*Mechanical properties and microstructure of plain carbon steel.*
Not to scale.

The following groups, classified according to applications, are common in engineering and include:

(*i*) *Structural steels.* These should comply with BS 4360 : 1972 (Weldable structural steels) which gives requirements for structural steels including chemical composition and mechanical properties. Weather-resistant grades are also included (*see* Table LXXVIII).

Structural steel sections used in construction are covered in:

BS 4 : Part 1 1972 Hot-rolled sections.

Part 2 1969 Hot-rolled hollow sections.

Other relevant British Standards include:

BS 449 : Part 2 1969. The use of structural steel in building.

TABLE LXXVII: EFFECT OF ALLOYING ELEMENTS ON
THE PROPERTIES OF STEEL

Alloying element	General effect
Carbon (C)	● Forms interstitial solid solution with iron in steel ● Increase in strength, hardness and wear-resistance with increase in carbon content, but decrease in ductility and weldability ● Increase ductile/brittle transition temperature ● Delays the initiation of transformation to pearlite and bainite
Manganese (Mn)	● Goes into solid solution, improving strength and hardness, and also forms hard carbides (Mn_3C) ● Improves hardenability by lowering the critical cooling rate ● Improves ductility by increasing Mn content and decreasing C content for same strength ● Reduces brittleness due to sulphur (as MnS inclusions in steel) ● If excess present, can cause temper brittleness (which can be avoided by addition of Mo) ● Used as a deoxidiser in steel ● Lowers the eutectoid temperature
Nickel (Ni)	● Improves strength by going into solid solution and reduces the carbon content of the eutectoid (0·5% C eutectoid at about 10% Ni) ● Lowers the critical cooling rate, hence increasing hardenability ● Improves corrosion resistance to industrial atmospheres ● Improves resistance to fatigue ● Graphitising element (effect can be counter-acted by carbide-forming elements such as Mn, Cr or Mo) ● Stabilises austenite when present in sufficient amount
Chromium (Cr)	● Improves strength, hardness and hardenability ● Forms hard and stable carbides such as $(FeCr)_3C$, $(CrFe)_3C_2$, $(CrFe)_7C_3$, $(CrFe)_4C$ or Cr_4C

Alloying element	General effect
	● Goes into solid solution in ferrite or austenite
	● Improves resistance to corrosion, oxidation and wear
	Cr is a vital constituent in both heat-resisting and corrosion-resisting steels
Molybdenum (Mo)	● Dissolves in ferrite and in austenite to form solid solution
	● Forms complex carbides $(FeMo)_6C$, $Fe_{21}Mo_2C_6$, Mo_2C
	● Improves mechanical properties at high temperatures
	● Improves creep resistance at high temperatures
	● Reduces temper-brittleness in Ni–Cr steels
	Mo steels are characterised by high tensile and creep strength at elevated temperatures
Vanadium (V)	● Strong carbide-forming element like Mo
	● Grain-refining element
	● Inhibits grain growth at elevated temperatures
	● Strong deoxidiser
Silicon (Si)	● Dissolves in ferrite (maximum 18·5%) and in austenite (only 2%)
	● Strengthens ferrite
	● Improves hardenability
	● Improves resistance to corrosion and oxidation
	● A general purpose deoxidiser
Sulphur (S)	● Causes embrittlement (due to $Fe\,S$ formed)
	● Uniform distribution of $Mn\,S$ improves the free-cutting properties of steel
Tungsten (W)	● Dissolves in ferrite and austenite
	● Forms hard and stable carbides—WC, W_2C, Fe_3W_3C or Fe_4W_2C
	● Raises the critical temperature range
	● Refines grain size and produces less tendency to decarburisation during working
Titanium (Ti)	● Strong carbide-forming element
	2% Ti renders 0·50% carbon steel unhardenable

BS 970 :	Wrought steels in the form of blooms, billets, bars and forgings (5 Parts).
BS 1449 :	Steel plate, sheet and strip.
Part 1 1972.	Carbon steel plate, sheet and strip.
Part 4 1967.	Stainless and heat-resisting plate, sheet and strip.
BS 2691 : 1969	Steel wire for prestressed concrete.
BS 4461 : 1969	Cold-worked steel bars for the reinforcement of concrete (amended Feb. 1970).
BS 4483 : 1969	Steel fabric for the reinforcement of concrete (amended April 1972).

(ii) *Weathering steels.* These are low-alloy structural steels containing small amounts of additional alloying elements such as copper and chromium. They have better weather resistance to corrosion than mild steel. Two well-known ex-

TABLE LXXVIII: TYPICAL COMPOSITION AND MECHANICAL PROPERTIES OF STRUCTURAL STEELS

Material grade	Composition					Tensile strength (N/mm^2)	Yield strength (N/mm^2 min.)	Elongation (% min.)	Form
	C max.	Si	Mn max.	S max.	P max.				
40B[1]	0·25	—	1·6	0·06	0·06	400–480 400–480	230 240	25 25	Plates, sections and bars
40E[1]	0·19	0·10–0·55	1·6	0·05	0·05	400–480 400–480	260 255	25 25	Plates, sections and bars
Weathering grade									
WR 50A[1]	0·15C, 0·25–0·80Si, 0·70 max. Mn, 0·55 max. S, 0·065–0·160P, 0·30–1·25Cr, 0·25–0·55Cu					480 (min.)	340	21	Plates, sections and bars
Cor-ten A[2]	0·10C, 0·40Mn, 0·12P, 0·05S, 0·50Si, 1·0Cr, 0·02 min. Ti, balance Fe					480	340	19	Plates
Cor-ten B[2]	0·10–0·19C, 0·90–1·25Mn, 0·25–0·40Cu, 0·40–0·68Cr, 0·02–0·10V, balance Fe					480	340	19	Plates
Cor-ten C[2]	0·12–0·19C, 0·90–1·35Mn, 0·25–0·40Cu, 0·40–0·70Cr, 0·04–0·10V, balance Fe					550	410	16	Plates

[1] BS 4360 : 1972 Weldable structural steels.
[2] Woldhan, N. E. and Gibbons, R. C., eds., *Engineering Alloys*, Van Nostrand Reinhold Co., 1973.

amples are the *Cor-ten A and B* ("A" grade being cheaper than "B" grade). Typical compositions and mechanical properties are given in Table LXXVIII.

Weathering steels must be protected (*e.g.* by painting the surface) when used where sea-spray and permanently wet conditions can be damaging.

(*iii*) *Tool steels*. BS 4659 : 1971 (Tool steels) specifies requirements for six types of tool steel: high-speed, hot-work, cold-work, shock-resisting, special purpose and water-hardening. They must be hard and in the case of cutting tools must be able to retain hardness and cutting edge at high speed and at high temperature due to friction. They are all high-carbon steels, but high-speed tool steels contain in addition up to 22.5 per cent tungsten, 5 per cent chromium and 5·25 per cent vanadium.

Typical composition and mechanical properties of tool steels are given in Table LXXIX.

(*iv*) *Corrosion-resistant (or stainless) steels*. The remarkable resistance to corrosion found in stainless steels is due to the alloying element chromium, which forms a very adherent protective oxide film on the surface. Stainless steels are low-carbon steels containing at least 10 per cent chromium with or without other alloying elements (such as nickel, manganese, molybdenum, silicon, phosphorus, sulphur, nitrogen).

A very well-known type is the 18:8 chromium–nickel steel (BS types 302 and 304).

There are three main classes, according to the structure of the stainless steel:

I. *Martensitic* stainless steels are magnetic, the chromium content being limited to about 13 per cent. With more chromium, the steels become *ferritic* and when nickel is added they become *austenitic*. They can be hardened by heat treatment whereas ferritic and austenitic types cannot. However, austenitic stainless steels can be hardened by cold working. Martensitic types are more economical to produce, but are less resistant to corrosion.

II. *Ferritic* stainless steels contain a higher chromium content (13–27 per cent *Cr*). They cannot be hardened by heat treatment, are magnetic and are somewhat more corrosion-resistant than the martensitic steels. They are ductile and therefore can be drawn easily. The commonest one used in the building industry is the 17 per cent chromium steel (BS type 430 S15).

III. *Austenitic* stainless steels contain chromium and

TABLE LXXIX: TYPICAL COMPOSITION AND

Type	Composition (%)	Annealed hardness (HB max.)	Hardness (HV min.) after heat treatment	Heat treatment temperature range (°C)
High-speed (a) Molybdenum grades (BM) 6 grades	C 0·75–1·60 Si 0·40 max. Mn 0·40 max. Cr 3·75–5·0 Mo 2·75–10·0 W 1·0–7·0 V 1·0–5·25 Co 4·5–8·75	241–277	823–897	Annealing: 850–900 Preheating: 850 Hardening (salt bath): 1180–1235 Tempering: 520–570
(b) Tungsten grades (BT) 9 grades	C 0·60–1·40 Si 0·40 max. Mn 0·40 max. Cr 3·5–5·0 Mo up to 3·5 W 8·5–22·5 V 1·0–5·25 Co 0·60–12·25	255–302	798–912	Annealing: 850–900 Preheating: 850 Hardening (salt bath): 1220–1310 Tempering: 550–570
Hot-work a) Chromium grades (BH) 6 grades	C 0·30–0·45 Si 0·85–1·15 Mn 0·40 max. Cr 2·8–5·25 Mo 0·45–2·95 W up to 4·5 V 0·30–2·4 Co up to 4·5	229–248	–	Annealing: 850–870 Preheating: 800 Hardening (air or oil): 1000–1200 Tempering: 530–650
(b) Tungsten grades (BH)	C 0·25–0·60 Si 0·40 max. Mn 0·40 max. Cr 2·25–4·5 Mo 0·60 max. W 8·5–18·5 V up to 1·5 max. Co up to 0·60	235–241	763 (for BH 26)	Annealing: 870–890 Preheating: 800–850 Hardening (air or oil): 1100–1260 Tempering: 550–675
Cold-work (a) High-carbon high-Cr grades (BD) 3 grades	C 1·40–2·30 Si 0·60 max. Mn 0·60 max. Cr 11·5–13·0 Mo up to 1·20 V 0·25–1·00	225	735–763	Annealing: 850–870 Preheating: 800 Hardening (air or oil): 950–1030 Tempering ranges: 150–220 450–550

nickel as the main alloying elements. They are non-magnetic in the fully-annealed condition, but may become slightly magnetic during cold working. They have excellent corrosion resistance because of the higher chromium content. They are tough and ductile. The most important austenitic steels are the 18:8 (18 per cent Cr and 8 per cent Ni) and the 18:10:3 (18 per cent Cr, 10 per cent Ni and 3 per cent Mo) which are widely used in the fields of architecture and engineering.

PROPERTIES OF TOOL STEELS (DERIVED FROM BS 4659)

Type	Composition (%)		Annealed hardness (HB max.)	Hardness (HV min.) after heat treatment	Heat treatment temperature range (°C)
(b) Medium-alloy Air-hardening grades (BA) 2 grades	C Si Mn Cr Mo V	0·75–1·05 0·40 max. 0·30–2·1 0·85–5·25 0·90–1·6 up to 0·40 max.	241	735	Annealing: 730–870 Preheating: 650–800 Hardening (in air): 830–980 Tempering: 150–550
(c) Oil-hardening grades (BO) 2 grades	C Si Mn Cr W V	0·85–1·0 0·40 max. 1·1–1·8 up to 0·60 up to 0·60 0·25 max.	229	735	Annealing: 760–780 Hardening (in oil): 760–820 Tempering: 150–300
Shock-resisting (BS) 3 designations	C Si Mn Cr Mo W V	0·45–0·60 0·70–2·1 0·30–0·80 up to 1·7 up to 0·60 up to 2·5 0·10–0·30	229	600–655	Annealing: 790–820 Hardening (in oil or water): 870–950 Tempering: 175–650
Special-purpose (a) Low-alloy grade (BLS)	C Si Mn Cr V	0·95–1·05 0·40 max. 0·40 max. 1·3–1·5 0·10–0·30	207	760	Annealing: 790–810 Hardening (in oil or water): 790–840 Tempering: 150–350
(b) Carbon tungsten grade (BFI)	C Si Mn Cr W V	1·15–1·35 0·40 max. 0·40 max. 0·25–0·50 1·3–1·6 0·30 max.	207	760	Annealing: 780–800 Hardening (in oil or water): 780–800 Tempering: 200–250
Water-hardening (BW) 4 grades	C Si Mn Ni Cr Mo V	0·85–1·3 0·30 max. 0·35 max. 0·20 max. 0·15 max. 0·10 max. up to 0·35	207	790	Annealing: 740–790 Hardening (in water or brine): 770–790 Tempering: 180–350

Typical composition and mechanical properties of the three types of stainless steels are given in Table LXXX.

Relevant BSS include:

BS   470 : Part 4 1970 Wrought steels (blooms, billets, bars and forgings)—stainless, heat-resisting and valve steels.

BS 1449 : Part 4 1967 Steel plate, sheet and strip— stainless and heat-resisting plate, sheet and strip.

TABLE LXXX: TYPICAL COMPOSITION AND MECHANICAL PROPERTIES OF STAINLESS STEELS
(BS 970: Part 4: 1970)

Material (BS designation)	Composition	0.2% Proof stress (N/mm² min.)	Tensile strength (N/mm²)	Elongation (% min.)	Hardness (HB)
Martensitic steels					
420 S45	13 Cr, 0·32 C, others, balance Fe	495	700–850	15	201–255
431 S29	17 Cr, 2·5 Ni, 0·15 C, others, balance Fe	635	850–1000	11	248–302
Ferritic steels					
403 S17	13 Cr, 0·8 max. C, others, balance Fe	245	420 (min.)	20	170 (max.)
430 S15	17 Cr, 0·10 C, others, balance Fe	245	430 (min.)	20	170 (max.)
Austenitic steels					
316 S12	$Cr/Ni/Mo$ 17/12/2·5, 0·03 C, others, balance Fe	170	460 (min.)	40	183 (max.)
316 S16	$Cr/Ni/Mo$ 17/11/2·5, 0·07 C, others, balance Fe				
320 S17	$Cr/Ni/Mo$ 17/12/2·5 + Ti, 0·08 C, others, balance Fe	195	490 (min.)	40	183 (max.)

BS 1630 :	1957	13 per cent chromium steel castings for resistance to corrosion.
BS 1631 :	1957	Austenitic chromium–nickel steel castings for resistance to corrosion.
BS 1632 :	1957	Austenitic chromium–nickel–molybdenum steel castings for resistance to corrosion

(BSS 1630, 1631 and 1632 are included in BS 3100 : 1967).

| BS 3146 : | | Investment castings in metal. |
| | Part 2: 1975 | Corrosion and heat-resisting steels, nickel and cobalt base alloys. |

(v) *Heat-resisting steels* (BS 970 : Part 4 1970). At elevated temperatures, designers are faced with the problems of plastic deformation (creep) which occurs under prolonged loading and corrosion processes (*e.g.* by oxidation above a temperature of about 540 °C). It is found that by alloying steel with chromium and silicon resistance to corrosion by oxidation at high temperatures and to chemical attack by sulphur gases and flue dirt can be improved as a result of a hard protective film formed on the surface of the alloy. Nickel cannot be used in presence of sulphur dioxide (SO_2) and hydrogen sulphide (H_2S) gases. However, addition of aluminium with chromium and silicon can be used against attack by H_2S at 1000 °C. On the other hand, creep resistance can be improved by alloying additions of molybdenum, tungsten and vanadium.

Typical examples are:

Inconel (an 80–14–6 *Ni–Cr–Fe* alloy)—a scale-resisting steel often used for aero-engine exhaust manifolds;
Nimonic alloys (basically *Ni–Cr* alloys) often used for rotor blades of aero-gas turbines in the temperature range 700–1000 °C (*see* 11 on *Ni–Cr* alloys).

Typical composition and properties are given in Table LXXXI.

(vi) *Magnetic alloys.* Two types can be conveniently distinguished, according to whether they are magnetically "hard" or "soft." The magnetically "hard" type is used for permanent magnets while the "soft" type is used mainly for transformer cores, motor and generator armatures, etc. "Hard" types are characterised by high values of remanence

TABLE LXXXI: TYPICAL COMPOSITION AND PROPERTIES OF HEAT-RESISTING STEELS
(BS 970: PART 4: 1970)

Material (BS designation)	Composition	0.2% Proof stress (N/mm² min.)	Tensile strength (N/mm² min.)	Elongation (% min.)	Hardness (HB max.)
302 S25 (Austenitic)	Cr/Ni 18/9, 0.12 C, others, balance Fe	170	510	40	183
304 S15 (Austenitic)	Cr/Ni 18/9, 0.06 C, others, balance Fe	170	460	40	183
321 S12 (Austenitic)	Cr/Ni 18/9, 0.70 Ti, 0.08 C, others, balance Fe	195	490	40	183
410 S21 (Martensitic)	13 Cr, 0.12 C, others, balance Fe	340	550–700	20	152–207

and coercive force and usually contain alloying additions of chromium, tungsten and cobalt or of nickel, aluminium and iron (*e.g.* Alnico, Ticonal):

Alnico   (10Al, 18Ni, 12Co, 6Cu)
Ticonal   (8Al, 14Ni, 24Co, 3Cu).

"Soft" types should be readily demagnetised and are charac-. terised by high magnetic permeability. Typical examples are nickel–iron alloys such as Permalloy (78·5 per cent Ni, 21·5 per cent Fe) and Mu-metal (75 per cent Ni, 25 per cent Fe), which find wide applications in transformer cores. Relevant BSS include:

BS 1837 : 1970 Methods for the sampling of iron, steel, permanent magnet alloys and ferro-alloys.
BS 601   : 1973 Steel sheet and strip for magnetic circuits of electrical apparatus (five Parts).

COPPER AND COPPER ALLOYS

The predominant characteristic property of pure copper is its high electrical and thermal conductivity, which is higher than any other common metal (*see* Table LXX).

Copper is extracted from its ore copper pyrites (*see* Table LXIX) and refined copper is produced in various grades:

    (*i*) Cathode copper (BS 1035).
   (*ii*) Electrolytic tough-pitch high-conductivity copper (BS 1036).
  (*iii*) Fire-refined tough-pitch high-conductivity copper (BS 1037).
  (*iv*) Deoxidised copper (BSS 1172, 1174).
   (*v*) Oxygen-free copper (BS 1861).

3. Cu alloys (BSS 2870–2875). Important alloys are:

  (*a*) Brasses (*Cu–Zn*).
  (*b*) Bronzes (*Cu–Sn*), including gun-metals (*Cu–Sn–Zn*).
  (*c*) Aluminium bronzes (*Cu–Al*).
  (*d*) Cupronickels (*Cu–Ni*).

(*a*) *Brasses* (*see* Table LXXXII). Brasses are essentially alloys of copper and zinc, but other alloying elements such as lead, tin, iron, nickel, manganese and aluminium may be added to improve specific properties. They are non-magnetic. Tensile strength increases as the zinc content increases up to 45 per cent, and then decreases rapidly.

They form very useful engineering materials because of their wide range of mechanical properties, their ease of working and their high resistance to corrosion, in addition to

TABLE LXXXII: TYPICAL COMPOSITION AND PROPERTIES OF BRASS

Material designation	Composition	0·1% Proof stress (N/mm^2)	Tensile strength (N/mm^2)	Elonga- tion (%)	Hard- ness (HB)
CZ 125	95 Cu, 5 Zn	62	247	55	55
CZ 101	90 Cu, 10 Zn	62	262	55	60
CZ 102	85 Cu, 15 Zn	77	278	60	65
CZ 103	80 Cu, 20 Zn	108	293	65	70
CZ 105 CZ 106	70 Cu, 30 Zn	93	319	70	60
CZ 109	60 Cu, 40 Zn	123	355	45	90

NOTE: The values for mechanical properties which are dependent on the grain size relate to sheet and strip form in the annealed condition (unless stated otherwise).
Other forms also available are plate, rod, wire and tube.

their attractive appearance. Three important classes can be distinguished:

α brasses (0–38 per cent Zn).
αβ brasses (38–46 per cent Zn).
β brasses (46–50 per cent Zn).

(i) α brasses. These are malleable, ductile and relatively strong. They include:

Gilding metal, containing about 15 per cent Zn, often known as 85/15 brass.
Cartridge brass, containing 30 per cent Zn, known as 70/30 brass.
Admiralty brass, containing 70 per cent Cu, 29 per cent Zn and 1 per cent Sn. Similar to cartridge brass but has improved corrosion resistance due to the inclusion of tin.

(ii) αβ brasses. These are somewhat more brittle and therefore stronger owing to the higher zinc content. Muntz-metal (60 per cent Cu + 40 per cent Zn) is a typical example.

(*iii*) β brasses. These are hard and brittle—too brittle for most engineering purposes. Corrosion resistance is lower. A typical example is the brazing brass (50 per cent *Cu*, 50 per cent *Zn*, or 50/50 brass) which is useful for brazing, having a lower melting point than lower zinc brasses.

(*b*) *Bronzes* (*see* Table LXXXIII). Essentially alloys of

TABLE LXXXIII: TYPICAL COMPOSITION AND PROPERTIES OF BRONZE (SAND CAST) (BS 1400: 1973)

Material designation	*Composition*	*0·2% Proof stress* (N/mm^2)	*Tensile strength* (N/mm^2)	*Elonga- tion* (%)	*Hard- ness* (HB)
Admiralty gun-metal (G1)	88 *Cu*, 10 *Sn*, 2 *Zn*	130–160	270–340	13–25	70–95
Leaded gun-metal (LG2)	85 *Cu*, 5 *Sn*, 5 *Zn*, 5 *Pb*	100–130	200–270	13–25	65–75
Phosphor- bronze (PB4)	min. 9·5 *Sn*, max. 0·75 *Pb*, min. 0·4 *P*, max. 0·5 *Zn*, max. 0·5 *Ni*, balance *Cu*	100–160	190–270	3–12	70–95
Leaded phosphor- bronze (LPB1)	6·5–8·5 *Sn*, 2–5 *Pb*, min. 0·3 *P*, max. 2·0 *Zn*, max. 1·0 *Ni*, balance *Cu*	80–130	190–250	3–12	85–110

copper and tin, but other alloying additions may include zinc, phosphorus, nickel and lead in order to improve specific properties. They are generally harder, stronger but less ductile than the brasses.

They include:

(*i*) Coinage bronze (α bronze) (95 per cent *Cu*, 4 per cent *Sn*, 1 per cent *Zn*), used to be the standard British "copper" coinage, is soft and ductile.

(*ii*) Gun-metal (about 10 per cent *Sn*, 2 per cent *Zn*, balance *Cu*), very strong, hard and resistant to corrosion. Suitable for use as bearings, valve bodies and pipe unions.

(*iii*) Admiralty gun-metal (88 per cent *Cu*, 10 per cent *Sn*, 2 per cent *Zn*). Very strong and resistant to corrosion. Used for marine work.

(*iv*) Leaded gun-metals (85 per cent *Cu*, 5 per cent *Sn*, 5 per cent *Zn*, 5 per cent *Pb*). Soft due to inclusion of lead. More resistant to corrosion than the brasses.

(*v*) Phosphor-bronzes (10–14 per cent *Sn*, 0·1–0·3 per cent *P*). Addition of phosphorus increases strength and hardness and lowers ductility. Two types are distinguished:

I. Wrought phosphor-bronzes contain 3·0–8·5 per cent *Sn* and 0·1–0·3 per cent *P* and are used for springs and steam-turbine blading.

II. Cast phosphor-bronzes contain 9·0–13 per cent *Sn* and 0·3–1·0 per cent *P* and are used as bearing materials.

Typical composition and mechanical properties are given in Table LXXXIV.

(*vi*) Speculum metal (30–40 per cent *Sn*). Hard and brittle. High resistance to corrosion and wear.

TABLE LXXXIV: TYPICAL COMPOSITION AND MECHANICAL PROPERTIES OF PHOSPHOR-BRONZES

	Composition (percentage)			Tensile strength (N/mm²)	Elongation (%)
	Cu	*Sn*	*P*		
Wrought alloy	93·7	6	0·2	232 (cast)	18
				849 (drawn)	5
				370 (annealed)	65
Cast alloy	87·7	12	0·3	263	5

(*c*) *Aluminium bronzes.* These are copper-rich aluminium alloys containing no tin. They have a combination of properties such as high strength, good working properties, good resistance to corrosion, wear and fatigue, attractive golden colour and ability to undergo heat treatment similar to steels. Two important types are commonly used:

(*i*) Wrought α alloy contains 5–7 per cent *Al* and may be readily hot- or cold-worked. The important alloys are the 5 per cent *Al* alloy which is mainly used for decorative purposes and the 7 per cent *Al* alloy (also containing *Ni, Fe, Mn*) which is used for marine heat exchangers and desalination plant.

(ii) 10 per cent $\alpha\beta$ alloy contains 10 per cent Al with or without other alloying additions such as Fe, Ni and Mn and is used in situations where strength and good corrosion resistance are required.

Typical composition and properties are given in Table LXXXV.

TABLE LXXXV: TYPICAL COMPOSITION AND PROPERTIES
OF ALUMINIUM BRONZES

	Composition (%)		Tensile strength (N/mm²)	Elongation (%)
	Cu	Al		
Wrought α alloy	93	7	325 (cast)	69
10% αβ alloy	90	10	480 (cast)	20

(d) *Cupronickels*. These are alloys of copper and nickel and are extremely malleable. Tensile strength, hardness, ductility and corrosion resistance increase with the increase in nickel content. The useful alloys are the 80:20 and 70:30 cupronickels which are used mainly for condenser tubes and the 75:25 cupronickels used for the British "silver" coinage.

Typical properties of cupronickels in the annealed condition are tensile strength = 300–380 N/mm², percentage elongation = 40–45, Brinell hardness = 75–80.

ALUMINIUM AND ALUMINIUM ALLOYS

Aluminium and its alloys are widely used in the fields of engineering, automobiles and aircraft, due to their characteristic properties of lightness, high thermal and electrical conductivities and high corrosion resistance.

Pure aluminium (*see* Tables LXIX and LXX for extraction and properties) is soft, ductile and weak and therefore unsuitable for most engineering purposes. It can be strengthened by cold working and even more by alloying with other elements such as Mg or Si.

There are two major categories of aluminium alloys: those that can be strengthened by controlled heat treatment and those that can be strengthened solely by alloying and often cold working.

Aluminium alloys for general engineering purposes are covered by BSS 1470–75 (wrought forms) and BS 1490 (ingots and castings), and for aircraft materials by BS "L" series and DTD specifications (issued by the Directorate of Technical Development).

BS designations for a wide range of alloys are given in BSS 1470–75 and 1490 and should be consulted.

It is more convenient to subdivide aluminium alloys into:

(a) Wrought alloys:
- (i) Non-heat-treatable.
- (ii) Heat-treatable.

(b) Cast alloys:
- (i) Non-heat-treatable.
- (ii) Heat-treatable.

4. Wrought alloys (BSS 1470–1475). Heat-treatable alloys are designated in the BS by the prefix H and the non-heat-treatable ones by the prefix N. The form of material supplied is indicated by symbols *e.g.* S (sheet), P (plate), E (extruded bars and sections), T (tube), F (forgings), R (rivet stock), B (bolt and screw stock), V (extruded round tube and hollow sections), G (wire), C (clad sheet and strip), E (bar, rods and sections).

The common alloying elements are copper, silicon, manganese, magnesium and zinc—the total percentage often amounts to more than 10 per cent.

(a) *Non-heat-treatable alloys* cannot be strengthened by heat treatment but only by cold work. The main alloying elements are manganese (usually 1·25 per cent) and magnesium) 2–7 per cent). They have low to medium strength and are characterised by high resistance to corrosion and ease of working.

(b) *Heat-treatable alloys* can be hardened and strengthened by heat treatment and usually contain a larger number of alloying elements (such as copper, magnesium, manganese, silicon, zinc) than the non-heat-treatable alloys. The hardening medium is due to the formation of compounds such as Mg_2Si and $CuAl_2$. Three stages are involved in the heat treatment of aluminium alloys:

(*i*) *Solution treatment*, in which the alloy is heated between 450–550 °C until the alloying solute elements are completely dissolved.

(*ii*) *Quenching*, in which the solution-treated material is cooled rapidly to prevent the precipitation of the solute elements and to obtain a supersaturated solid solution.

(*iii*) *Aging*, in which the quenched alloy is left at ambient temperature for hardening to take place (known as *natural aging* which takes about four days or more for maximum hardness to develop). Hardening, however, can be accelerated by reheating the quenched alloy to a temperature between 110–215 °C for a few hours—this process is known as *artificial aging* or *precipitation hardening*.

Typical composition and properties of wrought aluminium alloys are given in Table LXXXVI.

5. Cast alloys (BS 1490) (prefixed LM). These are normally used for the various casting processes and are available in both the as-cast and heat-treated conditions. Properties of alloys are improved solely by alloying (non-heat-treatable cast alloys) and further improved by the effect of heat treatment (in the case of heat-treatable cast alloys).

The alloying elements are mainly copper, silicon, magnesium, zinc and iron—the total percentage seldom amounts to more than 10 per cent. Addition of copper, silicon, magnesium or manganese lowers the melting point of the cast alloy, and makes it more convenient for casting.

(*a*) *Non-heat-treatable alloys.* The important ones include:

(*i*) *Al–Si alloys* (*e.g. LM6* which contains 10–13 per cent *Si*) which are characterised by excellent fluidity, casting properties and good corrosion resistance.

(*ii*) *Al–Si–Cu alloys* (*e.g. LM2* which contains 10 per cent *Si* and 1·5 per cent *Cu*), in which the addition of copper increases the strength and machinability but lowers the castability, corrosion-resistance and ductility.

(*iii*) *Al–Mg–Mn alloys* (*e.g. LM5* which contains 4·5 per cent *Mg* and 0·5 per cent *Mn*), which are characterised by high corrosion-resistance, especially in marine environments.

(*b*) *Heat-treatable alloys.* Further improvement in strength by heat treatment is due to the formation of hardening media (*e.g. $CuAl_2$, Mg_2Al_3, $NiAl_3$*).

TABLE LXXXVI: TYPICAL COMPOSITION AND PROPERTIES OF WROUGHT ALUMINIUM ALLOYS (BS 1470: 1972 PLATE, SHEET AND STRIP)

Material designation	Composition (approximately)	0·2% Proof stress (N/mm² min.)	Tensile strength (N/mm²)	Elongation[1] (% min.)	Condition[2]
NS 3	Al–1 Mn, others: Cu, Mg, Si, Fe, Zn	— —	90–130 120–145	20 5	O H2
NS 4	Al–2 Mg, others: Cu, Si, Fe, Mn, Zn, Cr, Ti	60 130	160–200 200–240	18 4	O H3
NS 5	Al–3·5 Mg, others: Cu, Si, Fe, Mn, Zn, Cr, Ti	85 165	215–275 245–295	12 5	O H2
NS 8	Al–4·5 Mg, Mn, others: Cr, Si, Fe, Zn, Cr, Ti	125 235	275–350 310–375	12 5	O H2
HS 15	Al–4 Cu, Si, Mg, others: Si, Fe, Zn, Cr, Ti	245 375	385 (min.) 430 (min.)	13 6	TB TF
HS 30	Al–1 Si, Mg, Mn, others: Cu, Fe, Zn, Cr, Ti	— 120 255	155 (max.) 200 (min.) 295 (min.)	16 15 8	O TB TF

[1] Elongation on 50 mm material thicker than 0·5 mm.
[2] O = Annealed.
H2, H3 = Strain-hardened (H3 stronger than H2).
TB = Solution heat-treated only.
TF = Solution heat-treated and precipitation-hardened.

The important alloys include:

(i) *Al–Mg alloys* (e.g. *LM10* which contains 10 per cent Mg), in which the hardening medium Mg_2Al_3 is formed through the high magnesium content. The characteristic properties are high strength, good machinability and excellent corrosion-resistance.

(ii) *Al–Cu alloys* (e.g. *LM11* which contains 4·5 per cent Cu, and *LM14* which contains 4 per cent Cu, 2 per cent Ni, 1·5 per cent Mg), in which the hardening media are $CuAl_2$

and $NiAl_3$. This type of alloy has only moderate corrosion-resistance and poor casting characteristics.

Typical composition and properties are given in Table LXXXVII.

TABLE LXXXVII: TYPICAL COMPOSITION AND PROPERTIES OF CAST
ALUMINIUM ALLOYS (BS 1490: 1970)

Material designation	Composition	Tensile strength (N/mm² min.)	Elonga-tion (% min.)	Condi-tion[1]
Non-heat treatable:				
LM 6	Al–10·0/13·0 Si, others: $Cu, Mg, Fe, Ni,$ Zn, Pb, Sn, Ti	190 (chill-cast)	5	M
LM 2	Al–9·0/11·5 Si, 0·7/2·5 Cu, others: $Mg, Fe, Ni,$ Zn, Pb, Sn, Ti	150 (chill-cast)	—	M
LM 5	Al–3·0/6·0 Mg, 0·3/0·7 Mn, others: $Cu, Si, Fe, Ni,$ Zn, Pb, Sn, Ti	140 (sand-cast) 170 (chill-cast)	3 5	M
Heat-treatable:				
LM 10	Al–9·5/11·0 Mg, others: $Cu, Si, Fe, Mn,$ Ni, Zn, Pb, Sn, Ti	280 (sand-cast) 310 (chill-cast)	8 12	TB
LM 13	Al–0·7/1·5 Cu, 0·8/1·5 Mg, 10·0/12·0 Si others: $Fe, Mn, Ni, Zn,$ Pb, Sn, Ti	210 (chill-cast) 170 (sand-cast) 280 (chill-cast)	— — —	TE ⎱TF ⎰

[1] M = As cast.
  TB = Solution-treated.
  TE = Precipitation-treated.
  TF = Solution-treated and precipitation-treated.

MAGNESIUM AND MAGNESIUM ALLOYS

Magnesium is the lightest metal produced commercially (its specific gravity is 1·74 compared with 2·70 for aluminium and 7·80 for steel). Pure magnesium is not strong enough for engineering purposes, but the strength can be improved by alloying usually with elements such as manganese, aluminium, zinc, zirconium, etc.

The extraction and properties of magnesium are given in Tables LXIX and LXX.

There are four main classes of magnesium alloys:

6. Mg–Mn alloys. These are wrought alloys characterised by good corrosion-resistance and weldability.

7. Mg–Al–Zn alloys. These are available in the cast and the wrought forms. The cast alloys can be heat treated for increased strength and ductility.

8. Zirconium-containing alloys. Also available in the cast and the wrought forms. The addition of thorium and some of the rare earth elements (*e.g.* cerium) improves the strength characteristics and performance at elevated temperatures.

9. Mg–Li alloys. Containing 14 per cent lithium and 1 per cent aluminium, these are much lighter than pure magnesium and used in missiles and other aerospace vehicles.

Magnesium alloys are specified by the following BSS:

BS 2970 : 1972 Magnesium alloy ingots and castings.
BS 3370 : 1970 Wrought magnesium alloys for general engineering purposes—plate, sheet and strip.
BS 3372 : 1970 Wrought magnesium alloys for general engineering purposes—forgings and cast forging stock.
BS 3373 : 1970 Wrought magnesium alloys—bars, sections and tubes including extruded forging stock.

Typical composition and properties of magnesium alloys are given in Table LXXXVIII.

NICKEL AND NICKEL ALLOYS

Nickel is a ferromagnetic metal. It combines good strength properties at high and even sub-zero temperatures with good

TABLE LXXXVIII: TYPICAL COMPOSITION AND PROPERTIES OF MAGNESIUM ALLOYS (IN THE "AS MANUFACTURED" CONDITION)

Material BS[1]		Compositional designation	0.2% Proof stress (N/mm² min.)	Tensile strength (N/mm² min.)	Elonga- tion (% min.)
Mg–Mn alloys:					
	S		70	200	3–5
MAG-101	F	Mg–Mn 1·5	105	200	4
	E		120	230	4
Mg–Al–Zn alloys:					
MAG-111	S	Mg–Al 3 Zn 1 Mn	160	250	5–7
	E		160	245	10
Mg–Zn–Zr alloys:					
	S		160	250	5–6
MAG-151	F	Mg–Zn 3 Zr	180	270	7
	E		225	305	8

[1] BS 3370: 1970   S = Plate, sheet and strip.
BS 3372: 1970   F = Forgings and cast forging stock.
BS 3373: 1970   E = Extruded bars, sections and tubes.

corrosion resistance, particularly against caustic alkalis, ammonium salt solutions and organic acids. It is non-toxic and hence widely used in the manufacture of food-handling equipment. Extraction and properties of nickel are given in Tables LXIX and LXX.

A large number of nickel alloys are available for a wide range of applications. Important nickel-based alloys include:

10. Ni–Cu alloys. Nickel and copper are mutually soluble in all proportions to form Ni–Cu alloys, whose strengths increase with increasing nickel content up to 60–70 per cent Ni. They have very good corrosion-resistance, particularly against sea water, alkalis and certain acids.

Typical alloys are:

(a) Monel (a proprietary alloy containing approximately 68 per cent Ni and 32 per cent Cu with small amounts of Mn, Fe, Si and C): good mechanical properties at elevated temperature and excellent corrosion-resisting properties.

(b) K-Monel (a modification of Monel, obtained by addition of approximately 4 per cent *Al*): due to the addition of aluminium this alloy can be heat-treated further to improve the mechanical properties. Like Monel, it has excellent resistance to atmospheric and chemical corrosion.

(*c*) *Cupronickels* (*see* copper alloys, **3** (d)).

11. Ni-Cr alloys. These are characterised by their heat- and creep-resistance.

Typical examples include:

(*a*) *80/20 nickel-chromium alloys:* used mainly for electrical resistance purposes and heating elements.

(*b*) *Inconel* (a proprietary alloy containing approximately 77 per cent *Ni*, 15 per cent *Cr* and 8 per cent *Fe*): good corrosion resistance and good mechanical properties, even at elevated temperatures.

(*c*) *Nimonic alloys* (based on the 80/20 nickel-chromium alloys with small additions of cobalt, titanium, aluminium and carbon): High creep strength due to additions of titanium and aluminium, good corrosion-resisting and good mechanical properties at elevated temperatures.

12. Ni-Fe alloys. Depending on composition, have high magnetic permeability and low thermal coefficient of expansion.

Typical examples include:

(*a*) *Permalloy* (a proprietary alloy containing 78·5 per cent *Ni*, 21 per cent *Fe*).

(*b*) *Alnico alloys* (approximate composition: 18 per cent *Ni*, 10 per cent *Al*, 12 per cent *Co*, 6 per cent *Cu*, balance *Fe*)—used for the production of nickel-alloy permanent magnets.

Nickel alloys are covered by the following BSS:

BS 374: 1963 *Cu–Ni* sheet, strip and foil (withdrawn, superseded by BS 2870).
BS 3071: 1959 *Ni–Cu* alloy castings.
BSS 3072–3076: 1968 *Ni* and *Ni* alloys—sheet and plate, strips, tubes, wire, rods (amended Nov. 1971).

Typical composition and properties of nickel alloys are given in Table LXXXIX.

TABLE LXXXIX: TYPICAL COMPOSITION AND PROPERTIES OF
NICKEL ALLOYS

Material	Composition (%)	0·2% Proof stress (N/mm²)	Tensile strength (N/mm²)	Elonga- tion (%)	Hard- ness (HV)
Ni–Cu alloys:					
Monel	67Ni, 30Cu, + Mn, Fe, Si, C	170–310	480–590	50–30	100–140
K-Monel	Monel + 2–4% Al	680–900	960–1160	30–15	280–340
Ni–Cr alloys:					
Inconel	77 Ni, 15 Cr, 8 Fe	310–390	560–710	30–20	160–200
Nimonic 80A	54 Ni, 20 Cr, 20 Co, 2·4 Ti, 1·4 Al + Si, Mg, C		1065	2·5	
Ni–Fe alloys:					
Permalloy	78·5 Ni, 21·5 Fe		High permeability alloy		
Alnico	18 Ni, 10 Al, 12 Co, 6 Cu, small amount Ti, balance Fe		Permanent magnet alloy		

TIN AND TIN ALLOYS

Tin is one of the earliest metals, known to man since about 4000 B.C. (the Bronze Age). Pure tin is ductile, and soft. Its useful characteristics are non-toxicity, low melting point and high resistance to corrosion. Hence it is widely used as coatings (*e.g.* tinplating) in the food-handling industry, as solders (tin–lead alloys), as bearing materials and in the development of organo-tin compounds.

Extraction and properties of tin are given in Tables LXIX and LXX.

Important tin alloys include:

13. Solders. These are alloys of tin with various proportions of lead, which are low-melting. Soft solders widely used in electronic fields contain about 60 per cent tin and 40 per cent lead (low melting point close to the eutectic point at 183 °C). Tin–lead alloys containing antimony or silver have better high-temperature strength than lead–tin solders.

14. Bearing materials. These are tin alloys containing antimony (7–9 per cent) and copper (3–4 per cent), known as Babbitt alloys, which are extensively used as bearings for large marine diesel engines and generating plant.

15. Bronzes. *See* copper alloys.

Tin and tin alloys are covered by the following BSS:

BS 3252: 1960 Ingot tin.
BS 219 : 1959 Soft solders.
BS 441 : 1954 Rosin-cored solder wire.
BS 3332: 1961 White metal bearing alloy ingots.

Typical composition and properties of tin alloys are given in Table XC.

TABLE XC: TYPICAL COMPOSITION AND
PROPERTIES OF TIN ALLOYS

Material (BS)[1]	Composition (%)				Hard-ness (HV)	0·1% Proof stress (N/mm²)	
	Sn	Pb	Cu	Sb			
3332/1	89·3	—	3·2	7·5	27	2·5	⎫
3332/2	86·8	—	4·2	9	29	3·4	⎪
3332/3	81	4	5	10	32	4·4	White
3332/4	75	9·5	3·5	12	31	4·5	metal
3332/6	59	28	3	10	27	4·1	bearing
3332/7	11·3	74·5	0·7	13·5	26	3·7	alloys
3332/8	5	78·5	0·5	16	22	3·5	⎭

BS 3332: 1961 White metal bearing alloy ingots.

LEAD AND LEAD ALLOYS

Like tin, lead is one of the oldest metals known to man. It is soft, malleable and dense (specific gravity 11·3). It has a low melting point but a high corrosion-resistance. It is toxic. Its main use as pure metal in our modern age is for shielding radiation such as X-rays and Gamma-rays. Extraction and properties of lead are given in Tables LXIX and LXX.
Important lead alloys include:

16. Pb/Sn alloys. These are available as soft solders (*see* tin alloys): tinman's solder, composition 62 per cent tin + 38 per cent lead, melting point 183 °C; plumber's solder, composition 66 per cent lead + 34 per cent tin, solidification over a wide range of temperatures.

17. Pb/Sn/Sb alloys. These are available as bearing metals (*see* tin alloys). These alloys are hardened by addition of antimony and tin.

18. Pb/Sb alloys. Containing 7–12 per cent antimony, these are used for storage-battery grids.

19. Fusible alloys. These are low-melting lead–tin alloys, containing bismuth, cadmium, antimony and mercury, used in dentistry.

Lead and lead alloys are covered by the following BSS:

    BS 3908: Parts 1–15 1965–1972 Sampling and analysis.
    BS 4513: 1969 Lead bricks for radiation shielding.
    BS 334: 1934 Chemical lead (types A and B).
    BS 2878: 1968 Determination of lead in gasoline (gravi-
                  metric method).
    BS 3909: 1965 Ingot lead for radiation shielding.
    BS 1178: 1969 Milled lead sheet and strip for building
                  purposes.

ZINC AND ZINC ALLOYS

Zinc is characterised by its relatively low melting point (419·5 °C) and its corrosion-resistance in the atmosphere and in neutral and alkaline aqueous solutions. Consequently it is mainly used as a protective coating for iron and steel and also as a sacrificial anode. Zinc coatings may be applied by galvanising, sherardising, spraying and painting with zinc-rich paints.

Extraction and properties of zinc are given in Tables LXIX and LXX.

20. Zn alloys. The two main groups of zinc-based alloys are:

(*a*) *Brass* (*Cu–Zn* alloys)—*see* copper alloys (3(*a*)). Alloys containing 30–37 per cent *Zn* are both ductile and strong to enable cold deformation into sheets and wire. Alloys containing 37–45 per cent *Zn* are less plastic when cold. In alloys containing about 50 per cent *Zn*, both ductility and strength are impaired.

(*b*) *Die-casting alloys*. These are covered in BS 1004. The most commonly-used alloys contain aluminium, magnesium and copper and are known under the proprietary name of "Mazak" alloys. Typical composition and properties are shown in Table XCI.

TABLE XCI: TYPICAL COMPOSITION AND PROPERTIES
OF ZINC ALLOYS

	Composition (%)	Tensile strength (N/mm²)	Elonga- tion (%)	Hard- ness (HB)
Mazak 3	4·1% Al, 0·04% Mg, balance Zn	280	15·2	83
Mazak 5	4·1% Al, 1·0% Cu, 0·04% Mg, balance Zn	330	9·2	92

Zn alloys containing small additions of copper and titanium show better creep strength than unalloyed zinc. They are available in sheets and strips and are used for prefabricated roofing, ventilation ducting and various pressings.

Small additions of tin, cadmium and lead may be harmful, as intercrystalline corrosion may occur.

Zinc and zinc alloys are covered by the following BSS:

BS 1004: 1972 Zinc alloys for die casting and zinc alloy die castings.
CP 3001: 1955 Zinc alloy pressure die casting for engineering.
BS 3436: 1961 Ingot zinc.

TITANIUM AND TITANIUM ALLOYS

Titanium is one of the newer engineering metals. It is silvery-grey in appearance, rather light (specific gravity 4·51, improving its strength/weight ratio). Its strength and ductility are dependent on the purity of the metal, which has an excellent corrosion resistance. Extraction and properties of titanium are given in Tables LXIX and LXX.

Titanium has a hexagonal structure below 882 °C (known as α phase) and a body-centred cubic form (known as β phase) above this temperature.

21. Ti alloys. The main alloying element is aluminium, whose addition is usually limited to 6 per cent since embrittlement occurs above this amount. Other alloying elements include manganese, copper, tin, vanadium, molybdenum, silicon, zirconium.

Three main groups of titanium alloys may be distinguished:

(a) αTi alloys, which are characterised by their strength, particularly at high temperatures, but which are difficult to work.

(b) βTi alloys, which are less strong, easier to work but rather unstable at elevated temperatures.

(c) $\alpha\beta$ alloys, which have intermediate properties.

Titanium alloys are used in applications where high strength/weight ratio, good creep strength at elevated temperatures (up to 500 °C) and excellent corrosion-resistance are required, such as in the aerospace and chemical engineering industries.

Titanium and titanium alloys are covered by the following BSS:

Aerospace TA series (applications in aircraft).
CP 3003 (9): Lining of chemical vessels.

Typical composition and properties of titanium alloys are given in Table XCII.

TABLE XCII: TYPICAL COMPOSITION AND PROPERTIES
OF TITANIUM ALLOYS

Material	Composition	Condition	Proof 0·1% stress (N/mm²)	Tensile strength (N/mm²)	Elonga- tion (%)	Reduc- tion in area (%)
Ti 115	Comercially	Annealed	232	417	30	50
Ti 130	pure Ti	Annealed	309	525	28	50
Ti 155		Annealed	540	695	25	
α alloys:						
Ti 230	2·5 Cu	Annealed	463	618	25	40
Ti 317	5 Al, 2·5 Sn	Annealed	772	880	16	32
αβ alloys:						
Ti 318	6 Al, 4 V	Annealed	957	988	15	45
Ti 550	4 Al, 4 Mo, 2 Sn, 0·5 Si	Heat-treated and aged	1000	1158	10	25
Ti 679	11 Sn, 2·25 Al, 5 Zn, 1 Mo, 0·2 Si	Heat-treated and aged	1000	1112	18	40

MECHANICAL TESTING OF METALS

Mechanical tests are carried out in order to determine the mechanical properties of metals and alloys, normally by deformation or destruction.

22. The tensile test (BSS 18, 3688, 4759). This test enables tensile strength and ductility of the metal to be determined. Tensile strength or tensile stress (TS) may be defined as the ability to resist a pulling stress without rupture. The test is usually carried out in a Hounsfield *tensometer* using test pieces which are carefully machined to a standard size and shape (Fig. 42). Ductility is defined as the ability to undergo cold plastic deformation before tensile failure (*see also* **24**).

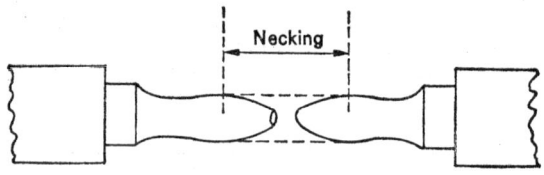

Fɪɢ. 42.—*Tensile specimen after fracture.*

(*i*) % *Elongation* (*El*) $= \dfrac{Increase\ in\ gauge\ length}{Original\ gauge\ length} \times 100$

(*ii*) % *Reduction in Area* (*RA*)
$$= \dfrac{Reduction\ in\ cross\text{-}section\ area}{Original\ cross\text{-}section\ area} \times 100$$

A typical tensile test curve is shown in Fig. 43(*a*). The test provides sufficient data for the following to be calculated: Yield stress, tensile strength, percentage elongation (per cent El), percentage reduction in area (per cent RA) and Young's modulus.

Where yield point is not so clearly shown, *e.g.* in non-ferrous metals and alloys, the 0·1 per cent or 0·2 per cent Proof Stress (PS) (*see* Fig. 43(*b*)) is calculated, thus:

$$0\text{·}1\ \text{per cent Proof Stress (PS)} = \dfrac{0\text{·}1\ \text{per cent Proof load}}{\text{Original cross-section area}}$$

Ductility of the test specimen may be deduced from the

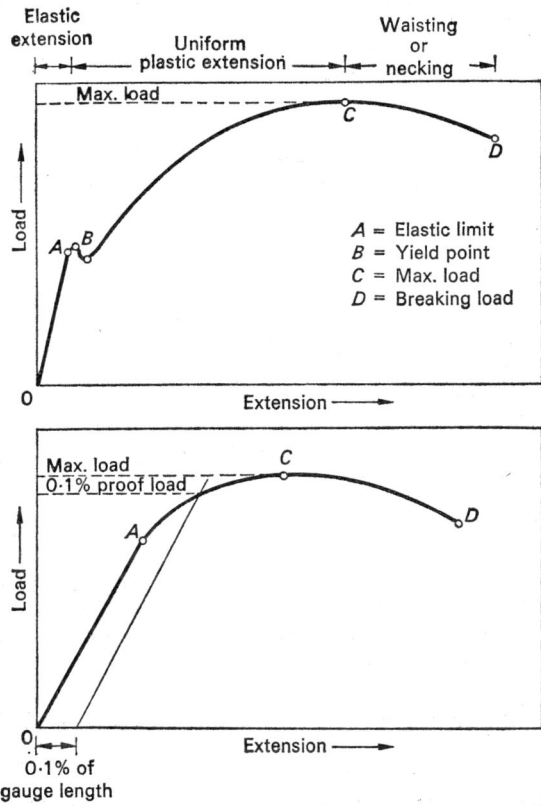

FIG. 43.—*Tensile test curves for metals.*

(a) (*top*) *Typical tensile test curve (yielding metals).*
(b) (*bottom*) *Typical tensile test curve (non-yielding metals).*

(i) *Yield point stress* $(YS) = \dfrac{Yield\ load}{Original\ cross\text{-}section\ area}$

(ii) *Tensile strength or*
    Ultimate tensile stress $(UTS) = \dfrac{Max.\ load}{Original\ cross\text{-}section\ area}$

(iii) *Young's modulus* $(E) = \dfrac{Stress}{Strain}$

$$= \frac{Load/Cross\text{-}section\ area}{Increase\ in\ gauge\ length/original\ gauge\ length}$$

shape of the load-extension curve and from the results of per cent El and per cent RA, for example:

(a) For non-ductile or brittle material: absence of plastic deformation prior to fracture, no per cent El and no per cent RA.

(b) Non-ductile material: uniform plastic deformation prior to fracture, low per cent El and low per cent RA.

(c) Ductile material: plastic deformation and necking prior to fracture, high per cent El and high per cent RA.

23. Hardness testing. Hardness is a complex property which is not simple to define. It may be taken as the resistance to indentation, abrasion, penetration, wear, machining, scratching, etc. The "scratch test" makes use of Moh's hardness scale (*see* Appendix II) which enables comparison of relative hardness of materials to be made.

More common tests such as Brinell, Rockwell, Vickers, Scleroscope, etc., make use of "indentation testing."

(a) *Brinell hardness test* (*BS 240*). This is based on the area of indentation made by pressing a standard hardened steel ball into the surface of the test specimen with a standard load (*P*) for a periodic time of 10–15 sec.

The hardness value, known as the Brinell hardness number (HB), is given by the relation:

$$\text{HB} = \frac{\text{Load}}{\text{Area of indentation}} = \frac{P}{\frac{\pi D (D - \sqrt{(D^2 - d^2)})}{2}} \text{ kgf/mm}^2$$

where P = load (kgf)—usually 3000, 1000 or 500 kgf.
D = diameter of the ball (mm)—usually 10 mm.
d = mean diameter of the indentation (mm) (measured by use of a microscope).

This test measures surface hardness only and is limited to materials less than HB = 480, as extremely hard materials deform the steel ball indentator. It cannot be used for testing materials at elevated temperatures.

(b) *Vickers hardness test* (*BS 427*). In this method, the spherical steel ball in the Brinell test is replaced by a harder indentator which is made of diamond and shaped in the form of a square-based pyramid with an angularity of 136 °. The load, which varies from 5 to 120 kgf, is applied usually for 15 sec. The Vickers hardness number (HV) or the Vickers pyramid number is given by the relation:

$$HV = \frac{\text{Load}}{\text{Area of indentation}}$$

$$= \frac{P}{d^2/(2 \operatorname{Sin} 68)} = 1 \cdot 854 \frac{P}{d^2} \text{ (approx.)}$$

where P = load (kgf).
d = mean of the 2 diagonals d_1 and d_2 (mm) (*see* Fig. 44).

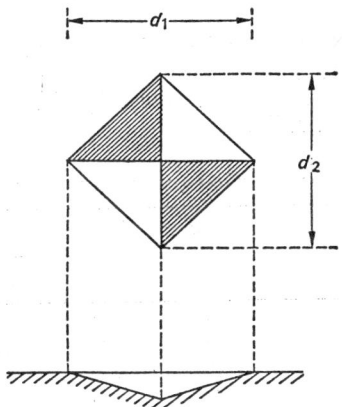

FIG. 44.—*Vickers diamond indentation.*

The hardness values are independent of the load, since the indentations are all geometrically similar. This test provides the same accuracy for thin, thick, hard or soft materials. Loading is automatic, which eliminates errors due to variations in rate, time and impact of loading.

(c) *Rockwell hardness test* (*BSS 891, 4175*). In contrast with the previous tests which measure the area of indentation, this Rockwell test measures the depth of indentation produced by a steel ball 1·58 mm diameter under a load of 100 kgf (Scale B) or by a diamond cone of angularity 120 ° under an impressed load of 150 kgf (Scale C).

The indentator is first loaded with a minor load of 10 kgf and applied to the test surface. The dial indicator (for measuring the depth of indentation) is set to zero. The load is then increased to 100 kgf (Scale B) or to 150 kgf (Scale C) and the depth of indentation measured by the dial gauge which gives a direct reading of Rockwell hardness number (HR). The relationship between HR and the depth increment (d mm) is as follows:

$$\text{HR (Scale C)} = 100 - \frac{d}{0 \cdot 002}$$

or

$$\text{HR (Scale B)} = 130 - \frac{d}{0 \cdot 002}$$

This test is simple and rapid and particularly useful for quality and process control.

Comparison of hardness values by the three methods is given in Table XCIII.

TABLE XCIII: COMPARISON OF HARDNESS VALUES

Material	HB	HV	HR	
			Scale C	Scale B
Soft brass	60	61	—	—
Mild steel	131	131	—	74
Soft chisel steel	235	235	22	99
White cast iron	415	437	44	114
Nitrided surface	745	1050	68	—

[Rollason, E. C., Metallurgy for engineers, Edward Arnold, 1973.

(d) *Impact testing* (*BS 131*). The impact test is designed to determine the resistance to suddenly applied stresses (BS 2094 definition). Various impact tests are available, using tension, torsion and bending techniques to provide a measure of toughness or impact-resistance.

The methods commonly used are based on the notched bar tests covered by BS 131:

Part 1: 1961 The Izod impact test on metals.
Part 2: 1972 The Charpy V-notch impact test on metals.

Part 3: 1972 The Charpy U-notch impact test on metals.
Part 4: 1972 Calibration of pendulum impact testing
machines for metals.
Part 5: 1965 Determination of crystallinity.

In the Izod test, a standard test piece (either 10 mm ×
10 mm square section or 11·43 mm diameter for circular
section) having a notch of specified dimensions is securely
clamped vertically in a vice (as shown in Fig. 45). The

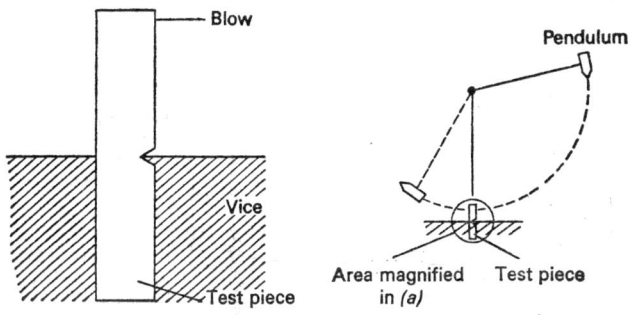

FIG. 45.—*Izod impact test.*
(*a*) (*left*) *Test piece.*
(*b*) (*right*) *Principle of test.*

striker or hammer attached to the end of a pendulum (Fig.
45(*b*)) is allowed to swing from a fixed height to hit the
protruding portion of the test piece. The difference between
the initial and final levels of pendulum swing gives a measure
of the energy absorbed in breaking the notched test piece.
This value (the Izod value) can be directly read from the
scale at the top of the pendulum, the higher the Izod value
the greater is the notch toughness or impact-resistance of the
material.

The Charpy test is based on a similar principle, by
measuring the energy absorbed in breaking by a single blow
from a pendulum test a notched test piece which, in this case,
is supported at each end.

24. Creep testing (BS 3500). Creep is defined as the deformation which proceeds slowly and continuously when stress is

applied at elevated temperatures (BS 2094). In steel, creep is negligible below about 300 °C.

Creep property, therefore, is time–temperature dependent. Creep testing is carried out by subjecting a test specimen to a constant load (usually in tension) and temperature (often elevated temperature) and measuring the extension produced

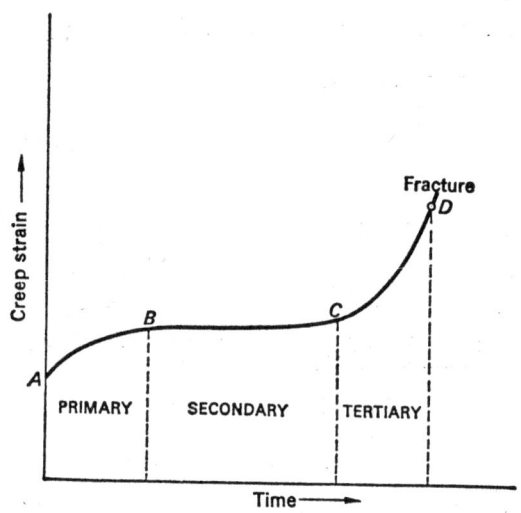

FIG. 46.—*Typical creep curve.*

at various time intervals until fracture occurs. A typical creep curve is shown in Fig. 46 which clearly defines three stages:

(a) *The primary or transient creep:* the extension is rapid at first, then slows down as it approaches the secondary stage.

(b) *The secondary creep or viscous flow (BC):* this is the most important part of the curve, where creep occurs at a more or less constant but minimum rate. It takes place over a long period of time.

(c) The *tertiary creep (CD):* this is when the rate of creep strain rises sharply, leading finally to fracture at D. The use of alloys in this tertiary stage should be avoided.

Creep testing is covered by BS 3500: Methods for creep and rupture testing of metals.

Part 1: 1969 Tensile rupture testing.
Part 3: 1969 Tensile creep testing.
Part 5: 1969 Production acceptance tests.
Part 6: 1969 Tensile stress relaxation testing.

25. Fatigue testing (BS 3518). The term "fatigue" refers to failure of a material when subjected to repeated cycles of applied stress. The stress required to cause failure in this way (*i.e.* under cyclic loading) is much less than that required to break the material under static loading.

There are various types of fatigue-testing machines, but the Wohler type is commonly used. In this test the test piece of diameter 6 mm is securely held in a chuck and rotated. The test pieces may be mounted as a cantilever with single-point or two-point loading or as a beam with four-point loading. The test piece may be fatigue-tested until failure occurs, or until a predetermined number of stress cycles is reached. A graph is plotted of the stress amplitude (S) against the logarithm of the number N of stress cycles to give the $S–N$ curve, shown in Fig. 47. Fracture will not occur below a certain stress, known as the *fatigue limit or endurance limit.*

Fatigue testing is covered by BS 3518: Methods of Fatigue Testing.

Part 1: 1962 General principles.
Part 2: 1962 Rotating bending fatigue tests (amended Dec. 1962).
Part 3: 1963 Direct stress fatigue tests.
Part 4: 1963 Torsional stress fatigue tests.
Part 5: 1966 Guide to the application of statistics.

NON-DESTRUCTIVE TESTING

Non-destructive testing makes use of measurements without damaging or destroying test objects. It is used to detect flaws or defects or to measure performance properties of engineering materials. Hardness testing (described in **23**) can be considered as a non-destructive testing.

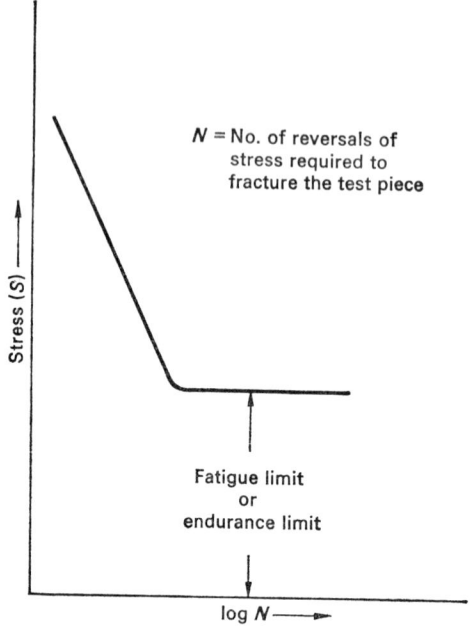

N = No. of reversals of
stress required to
fracture the test piece

Stress (*S*)

Fatigue limit
or
endurance limit

log *N*

FIG. 47.—*Typical S–N curve for a ferrous metal.*

Various methods are available for the non-destructive testing of metals:

    (*a*) Penetrant testing.
    (*b*) Magnetic particle testing.
    (*c*) Radiographic testing.
    (*d*) Ultrasonic testing.
    (*e*) Eddy current testing.

26. Penetrant testing. This is useful for detecting surface cracks or discontinuities. Test specimens, which should be clean, dry and free from scale, grease or other contaminants, are immersed or sprayed with a liquid penetrant which finds its way into cracks or discontinuities. The surface is rinsed to remove excess liquid, dried and then dusted with an absorbent powder such as chalk. Penetrant liquid trapped in cracks or discontinuities is absorbed by the fine powder and causes a stain characteristic of the nature of the imperfection. Such stain can be observed visually under normal or ultra-violet light (*see* Fig. 48). Dyes or fluorescent substances can be used in conjunction with the liquid penetrant to show up the stains more clearly. In a similar way, radioactive isotopes can be used in conjunction with a photographic film or a geiger counter.

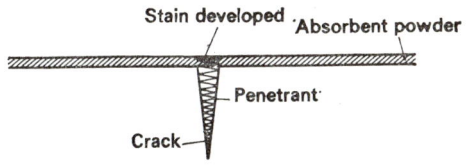

Fɪɢ. 48.—*Penetrant test.*

Penetrant testing is covered by the following BSS:
BS 3683: Glossary of terms used in non-destructive testing.
Part 1: 1963 Penetrant flaw detection.

BS 3889: Methods of non-destructive testing of pipes and tubes.
Part 3A: 1965 Penetrant testing of ferrous pipes and tubes.
BS 4080: 1966 Methods for non-destructive testing of steel castings.
BS 4124: Non-destructive testing of steel forgings.
Part 3: 1968 Penetrant flaw detection.
BS 4416: 1969 Methods for penetrant testing of welded or brazed joints in metals.

27. Magnetic particle testing. This is limited to ferromagnetic materials only, for detecting surface flaws. It consists of magnetising the test component or placing it in a strong magnetic field and applying a dry powder or a suspension of fine iron oxide in a liquid such as paraffin or kerosene. Patterns of magnetic flux can be observed and in the regions of flaws or cracks distortions in the pattern of magnetic flux are readily visible (*see* Fig. 49). This is a highly sensitive method, particularly for detecting cracks which are normal to the direction of magnetic flux. The use of fluorescent material together with the iron oxide powder will increase the sensitivity of the test when viewed under ultra-violet light.

The test specimen should be clean, dry and free from oil, grease, loose scale and dirt prior to testing. Often it may be necessary to demagnetise the specimen after test.

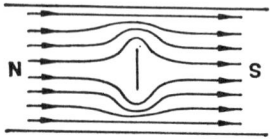 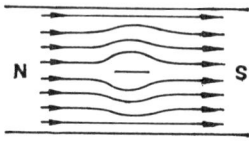

FIG. 49.—*Magnetic test.*

(a) (*left*) *Crack perpendicular to direction of magnetic flux.*
(b) (*right*) *Crack along the direction of magnetic flux.*

Magnetic particle testing is covered by the following BSS Specifications:

BS 3683: Glossary of terms used in non-destructive testing.

Part 2: 1963 Magnetic particle flaw detection (amended March 1968).

BS 3889: Methods of non-destructive testing of pipes and tubes.

Part 4A: 1965 Magnetic particle flaw detection: ferrous pipes and tubes.

BS 4080: 1966 Methods for non-destructive testing of steel castings.

BS 4124: Non-destructive testing of steel forgings.

Part 2: 1968 Magnetic particle flaw detection.

BS 4397: 1969 Methods for magnetic particle testing of welds.

28. Radiographic testing. In this test, use is made of electromagnetic radiations of very short wavelength (such as X-rays and gamma-rays) comparable to the inter-atomic spacings of materials under test to penetrate through the solid materials which are otherwise opaque to visible light. The intensity of the transmitted X-radiation after passing through the solid material can be measured on a photographic plate (film radiography) or on a phosphor fluorescent screen (fluoroscopy).

Imperfections will be shown up by shadow images on the film or screen.

In gamma radiography, the radioactive source used to emit gamma-rays may be radium, cobalt-60, caesium-137 or iridium-192. The technique is similar to X-ray radiography.

Radiographic testing is useful for irregular contours or rough surfaces but is expensive and requires stringent safety measures to avoid exposure of human bodies to radiation, since overexposure may lead to skin burns, ulcers and ultimately cancer.

Relevant BSS include:

BS 3683: Glossary of terms used in non-destructive testing.
Part 3: 1964 Radiological flaw detection (amended March 1968).
BS 4080: 1966 Methods for non-destructive testing of steel castings.
BS 2600: 1973 Methods for radiographic examination of fusion welded butt joints in steel.
BS 2910: 1973 Methods for radiographic examination of fusion welded circumferential butt joints in steel.
BS 2737: 1956 Terminology of internal defects in casting as revealed by radiography.
BS 3971: 1966 Image quality indicators for radiography and recommendations for their use.

29. Ultrasonic testing. Sound waves at high frequencies (between 20 kHz and 25 MHz) are transmitted through the material under test. The echoes which are reflected from the opposite surface boundaries or from imperfections can be received and displayed on a cathode-ray oscilloscope (CRO)—as shown in Fig. 50, the pulse-echo technique.

Other variations exist in ultrasonic test methods, such as:

(a) *Resonance testing*, in which the frequency of ultrasonic waves is adjusted to establish resonance with the material under test.

This enables the thickness and electrodynamic modulus of the material to be determined in addition to the detection of flaws within the material.

(b) *Measuring the transit time through the material* (see p. 58). Differences in transit time and hence velocity of the pulse through the solid material may also indicate differences in physical characteristics of the material.

Ultrasonic techniques are highly sensitive even for the detection of minute subsurface defects, but are limited to uniform sample geometry such as regular prisms or cylinders. Ultrasonic testing is covered by the following BSS:

BS 2704: 1966 Calibration blocks and recommendations for their use in ultrasonic flaw detection.

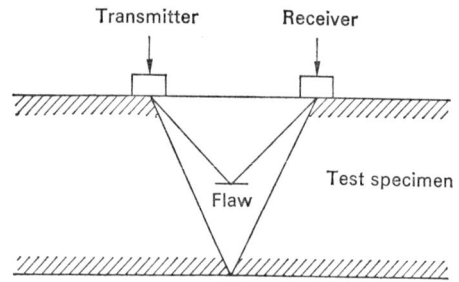

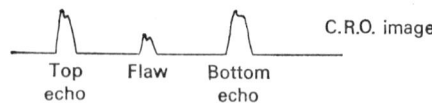

Fig. 50.—*Ultrasonic test (pulse-echo technique).*

(a) (*top*) *Principle.*
(b) (*bottom*) *C.R.O. image.*

BS 3683: Glossary of terms used in non-destructive testing.
    Part 4: 1965 Ultrasonic flaw detection (amended March 1968).
BS 3889: Methods of non-destructive testing of pipes and tubes.
    Part 1A: 1965 Ultrasonic testing of ferrous pipes (excluding casts) (amended July 1969).

BS 3293: Methods for ultrasonic examination of welds.
Part 1: 1968 Manual examination of fusion weld butt joints in ferritic steels.
Part 2: 1972 Automatic examination of fusion welded butt joints in ferritic steels.
Part 3: 1972 Manual examination of nozzle welds.
BS 4080: 1966 Methods for non-destructive testing of steel castings.
BS 4124: Non-destructive testing of steel forgings.
Part 1: 1967 Ultrasonic flaw detection.
BS 4331: Methods for assessing the performance characteristics of ultrasonic flaw detection equipment.
Part 1: 1968 Overall performance.
Part 2: 1972 Electrical performance.
BS 4336: Methods for non-destructive testing of plate material.
Part 1A: 1968 Ultrasonic detection of laminar imperfections in ferrous wrought plate.

30. Eddy-current testing. In this test, the specimen is placed inside a coil carrying an alternating current. A local current, known as eddy current, is induced within the conducting specimen. Interactions between the induced magnetic field of the specimen and the existing electromagnetic field of the current-carrying coil are observed or detected by a detector circuit. This enables imperfections such as cracks, seams, laps, wall-thinning, etc. to be detected.

This test is extremely rapid and is carried out without contact with the test specimen, but its use is limited to detection of surface and subsurface defects only. It cannot be used to detect defects at the centre of cylinders or rods.

Relevant BSS include:
BS 3683: Glossary of terms used in non-destructive testing.
Part 5: 1965 Eddy current flaw detection.
BS 3889: Methods of non-destructive testing of pipes and tubes.
Part 2A: 1965 Eddy current testing of ferrous pipes and tubes (amended July 1969).
Part 2B: 1966 Eddy current testing of non-ferrous tubes.

PROGRESS TEST 12

1. Which important metals are commonly used in engineering? (p. 239)

2. Outline two basic methods of extraction of metals from their ores. Illustrate by giving an example in each case. (p. 239)

3. Write down the carbon content in wrought iron, steel and cast iron. (**1, 2**)

4. How may plain carbon steels be classified? (**2**)

5. What are the effects of the following alloying elements on steel properties: (a) manganese, (b) nickel, (c) chromium, (d) silicon? (Table LXXVII)

6. What are stainless steels, and how may they be classified? (**2(b)**)

7. What are the characteristic properties of: weathering steels, tool steels and heat-resisting steels? (**2(b)**)

8. Show the steps involved in the extraction of copper from its ore. What are the various grades of refined copper commercially available? (Table LXIX)

9. What is the basic composition of brass? Distinguish between the three classes of brasses with reference to composition and mechanical properties. (**3(a)**)

10. What is the basic composition of bronze? Distinguish between the different classes of bronzes in terms of composition and mechanical properties. (**3(b)**)

11. Show the steps involved in the extraction of aluminium from its ore. (Table LXIX)

12. What is the effect of each of the following alloying elements on the properties of aluminium alloys: (a) copper, (b) silicon, (c) manganese, (d) magnesium, (e) zinc. (pp. 267–71)

13. Describe the principle of the heat treatment of aluminium alloys. (**4**)

14. What are the main classes of magnesium alloys? (**6–9**)

15. Write down typical compositions and mechanical properties of some important nickel alloys. (**10–12**)

16. Mention some of the important alloying elements of tin. Write down typical compositions of solders and bearing metals. (**13, 14**)

17. Outline the extraction of zinc from its ore. Mention some of the important alloying elements of zinc. (Table LXIX, **20**)

18. Give typical compositions and properties of important titanium alloys. (**21**)

19. Sketch a typical tensile curve for (a) a yielding metal, (b) a non-yielding metal.

Show how the following properties can be calculated: Yield stress, tensile stress, per cent elongation, per cent reduction in area and proof stress. (**22**)

20. What is meant by "hardness" of a material? Describe the principle of the following hardness tests: Brinell, Vickers and Rockwell. (23)

21. Describe a method of determining the impact strength of metals. (23)

22. What is meant by "creep?" How can it be measured? Sketch a typical creep curve. (24)

23. What is meant by "metal fatigue?" Describe a method of fatigue-testing in the laboratory and show a typical fatigue curve for a ferrous metal. (25)

24. Describe the principle of each of the following non-destructive tests on metals: (a) penetrant testing, (b) magnetic particle testing, (c) radiographic testing, (d) ultrasonic testing, (e) eddy-current testing. (26–30)

EXAMINATION QUESTIONS

1. Give *one-sentence answers* to the following questions:
   (a) Why is aluminium ductile?
   (b) Why is glass brittle?
   (c) Why is the ductile–brittle transition temperature usually lower in a notched tensile specimen test than in an impact specimen test?
   (d) Why is a high-carbon steel more prone to brittle fracture than a low-carbon steel?
   (e) What approximate magnification in hardness can be achieved by rapidly quenching a eutectoid steel from 800 °C.
   (f) What are the units of hardness?
   (g) In the equation:
      Diamond Pyramid Hardness $= C \times$ tensile strength
      what is the approximate value of the constant C?
   (h) What is the physical argument that is the basis of the relationship between hardness and tensile strength?
                                                    (C.E.I. Part 2)

2. In the construction industry the alloys of a metal are normally used much more extensively than the pure form of the metal. Discuss the reasons for this, giving examples where appropriate and citing any exceptions.
                                                    (I.O.B Ass. Part 1)

3. (a) What is meant by the following terms:
       (i) yield strength,
       (ii) tensile strength,
       (iii) ductility,
       (iv) toughness.

(b) Describe the tensile test and comment on its significance in the design of steel structures.

(c) Describe and account for the effect of carbon and manganese on the mechanical properties of structural steels.

(P.S.B. B.Sc. Bldg.)

4. (a) Describe briefly the mechanical test used to determine the following properties for a steel:

  (i) proof stress,
  (ii) ultimate tensile stress,
  (iii) elongation,
  (iv) reduction of area.

(b) Discuss the significance of the above properties in the design of steel structures.

(P.S.B. H.N.D. S.E.)

5. (a) Sketch and label the iron–iron carbide equilibrium diagram.

(b) Describe, with the aid of the diagram, the respective structural changes which occur when

  (i) 0·4 per cent carbon steel, and
  (ii) 1·2 per cent carbon steel are cooled slowly from 900 °C to room temperature.

(c) Compare and explain the difference in the mechanical properties of the two steels after

  (i) slow cooling, and
  (ii) rapid cooling.

(P.S.B. B.Sc. S.E.)

6. What are the conditions under which aluminium corrodes? Describe the engineering applications of aluiminium alloys.

(S.O.E. Grad. Exam. C.E.)

7. Explain what is meant by the stiffness of a testing machine and give examples of "hard" and "soft" systems.

What important parts of a tensile stress–strain curve for mild steel and a compressive curve for concrete are completely dependent on the stiffness of the testing machine; explain how with the aid of suitable illustrations.

(C.E.I. Part 2)

8. For what purposes is the indentation hardness of a material measured and to what extent is it related to other mechanical properties? Explain the principles of *one* of the following methods of hardness test:

  (a) Brinell,
  (b) Rockwell,
  (c) Vickers.

indicating clearly any advantages or limitations.

(C.E.I. Part 2)

FURTHER READING

Rollason, E. C., *Metallurgy for engineers*, Edward Arnold, 1973.
Bailey, F. W. J., *Fundamentals of engineering metallurgy and materials*, Cassell, 1972.
Wilson, C., *Design Engineering Handbook: Metals*, Product Journals Ltd., 1968.

HINTS ON ANSWERING QUESTIONS

Having studied a given section of the book, attempt to answer the questions in the Progress Test and Examination Questions, using the guidelines given below. Many of the questions give hints as to where the answers may be found, but you are advised to try to recall the information first, and only use the "hints" when absolutely necessary.

1. Read the whole question; if necessary re-read the question until you are clear as to what is being asked.
2. Prepare "keywords" for the various sections of your answer.
3. The answers (to these sections) will not necessarily be all of the same length. Devote more time to those sections you consider the more important.
4. Attempt to put down the required information using as few words as possible, avoiding padding and repetition.
5. Model answers should as far as possible consist of complete sentences.
6. Some answers may be improved (and often shortened) by means of suitable sketches.
7. Answers to questions involving calculations may often be checked by applying the test "Is the answer reasonable?"
8. Always read through your complete answer to check that nothing important has been omitted.
9. In answering examination questions, always try to find out what the examiner has in mind and keep as far as possible to the time allocated for each question. Remember, if you are asked to attempt, for example, five questions, always try to answer the full five questions.
10. Always do your best at the examinations and never allow your nerves get on top of you.

MOH'S HARDNESS SCALE

Material	Composition	Moh's hardness number
Diamond	C	10
Corundum	Al_2O_3	9
Topaz	$SiAl_2F_2O_4$	8
Quartz	SiO_2	7
Orthoclase	$KAlSi_3O_8$	6
Apatite	$Ca_5P_3O_{12}F$	5
Fluorite	CaF_2	4
Calcite	$CaCO_3$	3
Gypsum	$CaSO_2(OH)_4$	2
Talc	$Mg_3S_4O_{10}(OH)_2$	1

APPENDIX III

HEAT TREATMENT OF CARBON STEEL (0·1–1·7 %C)

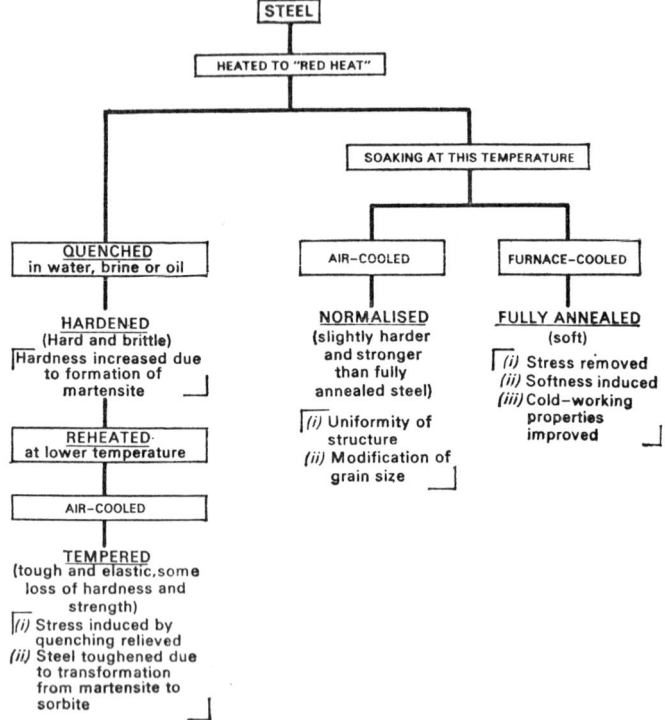

300

IRON–IRON CARBIDE EQUILIBRIUM DIAGRAM

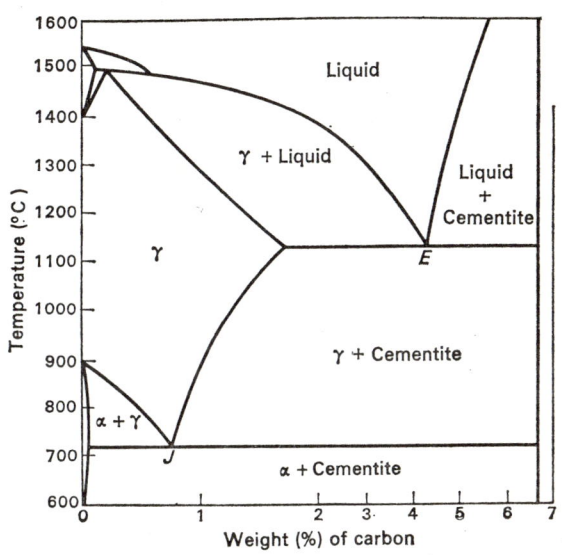

Approximately to scale.

γ = Austenite (solid solution of carbon in f.c.c. γ iron).
α = Ferrite (solid solution of carbon in b.c.c. α iron).
Cementite = Carbide Fe_3C.
E = Eutectic point.
J = Eutectoid.

DESIGN OF NORMAL CONCRETE MIXES (D.O.E. 1975)

The method of the *Design of Normal Concrete Mixes* by D. C. Teychenné, R. E. Franklin and H. C. Erntroy now replaces the method of Road Note No. 4.

The main reasons for this revision arise from the policy of metrication of the construction industry, the better understanding of the behaviour of concrete mixes, and the need for updating various data and relationships.

The important changes introduced are:

(*a*) The concept of *characteristic strength* and the *strength margin* according to CP 110.

(*b*) *Free water content* available for the hydration of the cement and for the workability of the fresh concrete.

(*c*) *Workability tests* include slump test for higher workability mixes and the V–B consistometer test for low workability mixes which require compaction by vibration. Since consistent relationships between the compacting factor and the slump or V–B tests are difficult to reproduce, the compacting factor test is not used in this method.

(*d*) *Types of aggregates*. Coarse aggregates are divided basically into crushed and uncrushed aggregates. Crushed aggregates are generally angular in shape and rough in surface texture, whereas uncrushed aggregates are generally smooth-textured but may be rounded or irregular in shape. Use of crushed aggregate results in a concrete of lower workability but higher strength.

(*e*) *Aggregate grading*. The concept of the fine aggregate grading zones specified in BS 882 is used to replace the earlier method of combined aggregate grading curves.

(*f*) *Mix parameters* are specified in terms of the weights of the ingredients in a unit volume of fully compacted concrete instead of the system of proportions or ratios.

The process of the method is divided into five stages, each of which deals with a particular aspect of the design and ends with an important mix parameter or the final unit proportions:

Stage 1: Strength → free-water/cement ratio
Stage 2: Workability → free-water *content*
Stage 3: Stage 1 and Stage 2 → cement content

Stage 4: Determination of the total aggregate content
Stage 5: Determination of the fine and coarse aggregate contents

Trial mixes: Trial mixes are then made, tested, checked and adjusted as appropriate.

The present revised method of mix design is applicable to concrete for most purposes, including pavements. It is also applicable to concrete mixes containing air-entraining agents as well as to concrete mixes for indirect-tensile strength. However, there are restrictions to this method, which makes reference only to the use of Portland cements such as OPC, RHPC or SRC and of natural aggregates or coarse air-cooled slag.

For details of the method, the reader should refer to the publication *Design of Normal Concrete Mixes* (H.M.S.O., 1975).

INDEX

305